THE ROUGH GUIDE to The Earth

www.roughguides.com

The Rough Guide to the Earth

Editing, typesetting and picture research: Ruth Tidball
Design: Duncan Clark
Diagrams: Ed Wright
Proofreading: Wendy Smith
Production: Aimee Hampson and Katherine Owers

Rough Guides Reference

Series editor: Mark Ellingham
Editors: Peter Buckley, Duncan Clark, Tracy Hopkins, Sean Mahoney, Matthew Milton, Joe Staines, Ruth Tidball
Director: Andrew Lockett

Front cover image: Earth with Pacific rim and outer space © Donald E. Carroll/Getty Images
Back cover image: lava entering ocean, Kilauea Volcano, Hawaii © G. Brad Lewis/Getty Images

This first edition published April 2007 by
Rough Guides Ltd, 80 Strand, London WC2R 0RL
345 Hudson St, 4th Floor, New York 10014, USA
Email: mail@roughguides.com

Distributed by the Penguin Group
Penguin Books Ltd, 80 Strand, London WC2R 0RL
Penguin Putnam, Inc., 375 Hudson Street, NY 10014, USA
Penguin Group (Australia), 250 Camberwell Road, Camberwell, Victoria 3124, Australia
Penguin Books Canada Ltd, 90 Eglinton Avenue East, Toronto, Ontario, Canada M4P 2YE
Penguin Group (New Zealand), 67 Apollo Drive, Mairongi Bay, Auckland 1310, New Zealand

Printed in Italy by LegoPrint S.p.A

Typeset in DIN, Myriad and Minion

304 pages; includes index

A catalogue record for this book is available from the British Library

ISBN 13: 9-781-84353-589-8
ISBN 10: 1-84353-589-0

1 3 5 7 9 8 6 4 2

THE ROUGH GUIDE to

The Earth

by

Martin Ince

www.roughguides.com

Acknowledgements

I would like to thank Andrew Lockett of Rough Guides for commissioning this book and for his thoughtful guidance throughout its writing. And I am especially grateful to Ruth Tidball, development editor at Rough Guides and much the most thorough book editor I have ever worked with, for her unstinting efforts to improve it.

About the author

Martin Ince is a science journalist based in London. He has written for many newspapers and magazines, especially *The Times Higher Education Supplement*, where he was deputy editor. He works as a media adviser and trainer for British and other European research organizations. This is his ninth book.

Contents

Introduction

This book will tell you about the Earth, an endlessly interesting corner of the universe that we call home.

Many years ago, a university handed me a degree certificate saying that I had studied geology. But I have spent the intervening period simplifying and explaining science, not doing it. So you will have no trouble understanding what follows. My aim, as Albert Einstein put it, is to provide explanations that are as simple as possible, but no simpler. And there will be some good pictures, plus some ideas about places you might want to take a look at for yourself.

People have studied the Earth for thousands of years. At first many of the questions they asked involved matters of life and death. When should I plant my spring crop? When will there be a high tide to launch my fishing boat? How can this village get enough drinking water? People still worry about problems such as these. But our knowledge of the Earth as a whole has grown beyond measure in the recent past. In these pages I will say a little about how we know what we know, and who first discovered it, as well as telling you what we have found out.

For most of the hundreds of thousands of years that humans have existed, they knew little about the planet they lived on. It was only in about 200 BC that the first measurement of the size of the Earth was made, the first vital step in understanding its true nature. Only much more recently, in the past few hundred years, have we learned the other key facts about it. We know how old the Earth is, what it is made of, what the inaccessible parts thousands of kilometres below our feet are like, how life has changed here over time, and how the solid, liquid and airy parts of it move over different timescales. And our new knowledge of other planets is allowing us to see the Earth in its true context for the first time.

Despite our learning, the Earth throws up mysteries. Some important puzzles have evaded solution for decades. We know what causes earthquakes. But we have only very limited abilities to predict them, despite big spending on research in earthquake-prone parts of the world such as Japan and California.

In the chapters that follow, I will try to explain what we know about where the Earth came from, how it fits into the universe around it, and what it consists of. This will involve looking at mountains and continents, seas and rivers, snowfields and glaciers, and wind and weather.

We will see how the deep Earth affects things at the surface, for example by generating the Earth's magnetic field. We will find out that despite its apparently unchanging face, the Earth is a place where continents move, volcanoes appear and meteorite impacts alter the environment.

And towards the end of the book, we will take a look at the way human beings are altering the Earth. The technology we command, the demands we make and the speed with which we do things all go far beyond anything the Earth is used to. How much difference are we making to the Earth, and what should we do about it?

This message about the demands we are making on this medium-sized planet may seem solemn. But I also hope to show that our knowledge of the Earth is very complete and satisfying. It provides the basis for a considered approach to managing and conserving the Earth that is now beginning to emerge.

Martin Ince
roughguide@martinince.com
London, 2007

1

Planet Earth

1

Planet Earth

The answer to the question "What is the Earth?" might seem obvious: it is a planet. But just what is a planet? How are planets formed? What distinguishes them from other objects in the universe? And, perhaps most importantly for us, how unusual a planet is Earth – what characteristics does it share with its cousins, and what makes it unique?

What is a planet?

The basic definition of a planet is simple enough. A planet is too small and cool to be a star, but massive enough to form a solid globe. Many, such as the big outer planets of our solar system, hide this solid surface beneath a dense atmosphere. All the planets we know are in orbit around a star; however, there could be some that are not. It is possible for gravitational forces in star clusters to fling planets into the cold of deep space, far from any warming stars, where they would be exceptionally hard to detect. We also know of some planet-sized satellites in our own solar system. For instance, Triton, the biggest satellite of Neptune, is bigger than Pluto. But if you haven't got a badge, you aren't a cop, and if you are orbiting a planet, you can't be a planet yourself.

Studying the solar system

The story of our understanding of planets and the Earth's place among them begins with the **ancient Greeks**. The word "planet" has its origins in the Greek verb meaning "to wander". Like all ancient cultures, the ancient Greeks spotted that, in addition to the Moon, there were five objects in the night sky that moved about. They were the planets that we call **Mercury**, **Venus**, **Mars**, **Jupiter** and **Saturn**.

If you spend a night staring at the sky, doing it without a telescope like our ancestors had to, you will notice that very little happens. Stars rise, wheel across the sky and set. Come back the next night, and you will see pretty much the same thing, although everything will happen a few minutes differently from the day before until it is back in the same place a year later. This may seem so dull that it is a wonder astronomy ever got going as an occupation for the lively-minded.

It is these five planets that provide most of the variety we see in the naked-eye sky. They come and go, they vary in brightness, and they move about. One, Mercury, is never very far from the Sun in the sky. Another, Venus, also stays near the Sun but is less tightly bound. The other three can be seen in the full dark of night. They move across the sky – Mars fastest, then Jupiter, then Saturn – in patterns which gave ancient thinkers pause for thought. Like Mercury and Venus, they chase around the same broad path across the sky as the Sun and Moon, an area called the **Zodiac**, which has the acute drawback of permitting the development of astrology. Their pace varies and sometimes when they are at their brightest, they turn round and loop back for a bit before resuming their forward progress.

But what had these lights moving through the night sky got to do with the Earth beneath our feet? The discovery that these strange movements could be explained was one of the greatest human achievements. It was made mainly by a Pole, **Nicolaus Copernicus** (1473–1543), and a German, **Johannes Kepler** (1571–1630), who realized that a simple change in assumptions could make everything clear. The new assumption was that the Earth was a planet just like the others and that they were all in motion around the Sun. This belief had been on the scene since the time of the ancient Greeks, but its proof is one of the great foundation stones of modern science. (No theory I have heard of has ever seriously questioned that the Moon orbits the Earth.)

The idea that the Earth is just one of a number of planets orbiting the Sun is a very fundamental and unsettling one. It removes the Earth and, more importantly, the human race from the centre of the universe. The Christian Church once found this hard to cope with. Today most Christians regard our knowledge of the universe as an affirmation of their faith rather than an attack on it.

Around this time, we also gained our first definite knowledge of other planets. The first person to look at the sky through a telescope was an Englishman, **Leonard Digges**, in about 1571. His work did not lead to

Flat Earth?

Portuguese explorer Ferdinand Magellan (1480-1521) led the first expedition to circumnavigate the Earth (although he died on the journey). He said he knew the Earth was round because he had seen its shadow on the Moon, during a lunar eclipse. He added that although the Church said the Earth was flat, he would rather trust a shadow than the Church.

But in fact belief in a flat Earth had never been an essential part of the Christian faith, and the Church did not regard Columbus's attempt to get to Asia by sailing west as a challenge to its authority. The common misconception that until the age of exploration people – and the Catholic Church in particular – believed the Earth was flat has its roots in Washington Irving's *The Life And Voyages Of Christopher Columbus*, published in 1828. In fact, 1700 years before Magellan, Eratosthenes (276–195 BC) had produced a pretty accurate estimate of the circumference of the Earth, and by the time of Pliny the Elder (1st century AD), it was generally accepted that the Earth was a sphere. In the Middle Ages, theologians engaged in vigorous debate about whether the antipodes were inhabited, or even reachable, but any doubts about the Earth's spherical shape had long been laid to rest in the minds of most educated people. As our knowledge of the rest of the universe grew, so did it become increasingly hard to believe that the whole lot could rotate round the Earth every 24 hours.

Images from spacecraft have now provided us with pictorial proof that the Earth is round. And there are plenty of other indirect proofs that the Earth is a sphere. One is that we don't see any other plate planets. Another is that the Earth's weather systems only make sense if the solid surface beneath them is a sphere. And exactly what is below if the Earth is flat? But in a spirit of openness, here is a website that will try to persuade you that the Earth is flat. It belongs to the Flat Earth Society. If it convinces you, stop reading this book right now, as everything in it is wrong.

The Flat Earth Society www.theflatearthsociety.org

radical new discoveries, although he did memorably compare the surface of the Moon to a pie that his cook had made. This was radical in its own way, since it broke with the accepted wisdom that all celestial objects were perfect because they were at a safe distance from sinful Earth.

Despite being armed with a crude telescope that was less powerful than a good pair of modern binoculars, **Galileo Galilei** (1564–1642) was much more successful. He discovered that Venus has phases like the Moon, showing different amounts of its lit face to us as its position in the sky changes relative to the Sun, that Jupiter has moons of its own, that great craters mark the Moon's surface and that Saturn, well, looks a bit odd. Later and better telescopes showed that this was because it is surrounded

The dullest debate in science

"I have found your Planet X", said Clyde Tombaugh to Vesto Slipher on 18 February 1930. Tombaugh was a humble observer, while Slipher was a top astronomer, and Tombaugh's boss at the Lowell Observatory in Arizona. But finding Pluto put Tombaugh into the record books and ensured that he is probably better known today than Slipher.

Back then, things were simple. You just needed to give the new planet a name from classical mythology, and announce that the solar system had grown a little. But in the twenty-first century, we have realized that Pluto is only one of the objects that make up the outer solar system. More tellingly for some, it is not even the biggest. The romantically named 2003 UB313 is about 2400km in diameter, compared to 2320km for Pluto, which is dramatically smaller than everyone thought in 1930.

These objects – primitive remnants of the early solar system – are the biggest known members of the **Kuiper Belt**, a zone beyond Neptune that may contain thousands of objects over 100km in diameter.

Our increasing knowledge of the outer solar system has led to a lengthy debate about whether Pluto deserves its planet status. Finally, in 2006, the International Astronomical Union decided that Pluto should be demoted to the status of "dwarf planet", along with UB313, now called Eris, Ceres, the biggest of the asteroids, and Charon, Pluto's own satellite. Perhaps this long-running discussion can now end.

by its now-famous rings. Galileo's discoveries showed that other planets are round objects a little like the Earth.

One idea that nobody really caught on to was the notion that there might be other planets beyond those known from the earliest times. In 1781, however, a German astronomer based in England, **William Herschel** (1738–1822), found the planet **Uranus** while conducting a complete survey of the sky visible from England. This was one of the most mind-changing discoveries in the history of science, showing that the fixed universe of the past was to be replaced in the imagination by a dynamic place where new types of objects might be found.

In 1846 the planet **Neptune** was discovered (Galileo had seen it in 1612 without realizing its importance), and in 1930, **Pluto**. While there is an ongoing debate (see box) about exactly what is and is not a planet in the outer reaches of the solar system, it is now well established that the Earth is one of a group of planets which resemble each other more or less closely and which, therefore, can help us to understand the Earth itself in more detail.

The solar system: an inventory

One star

Our star, the Sun, weighs about 2 x 10^{30} kg, an amount generally used as the yardstick for other stars (so Sirius, the brightest star in the night sky, is 2.14 solar masses, or 2.14 sols). It is 1.4 million km across, about 400 times the diameter of the Earth. Although the solar system contains many astounding objects, don't forget that the Sun contains over 99 percent of its matter. Its power output is about 4 x 10^{26} watts, about 40 million million times as much energy as humans use. This is one reason why solar power is probably worth more attention.

Eight planets and three dwarf planets

Working outwards from the Sun, here are the accepted planets and dwarf planets of the solar system as of 2007 with their vital statistics:

Name	*Diameter, km*	*Mass, Earth = 1*	*Average distance from Sun, million km*	*Time to orbit Sun*	*Satellites*
Mercury	4878	0.05	58	88 days	none
Venus	12,102	0.95	108	225 days	none
Earth	12,756	1	150	365 days	1
Mars	6780	0.11	228	687 days	2
Ceres	940	0.00016	414	4.6 years	none
Jupiter	143,000	318	781	11.86 years	100+?
Saturn	121,000	96	1426	29.5 years	30+
Uranus	51,100	14	2879	83.7 years	21+
Neptune	49,500	17	4500	165 years	13
Pluto	2390	0.002	5900	249 years	3
Charon	1205	0.0005	5900	249 years	none
Eris	2400	0.002	10,122	557 years	none?

Their satellites

The space probe era has brought us more information on the satellites of the solar system than a book about the Earth can cope with. The main facts are that: the Earth has one, the Moon; Mars has two which are just captured asteroids; the gas giants have them in several flavours; and Pluto and its satellite Charon are close to being a double planet, as Charon has 12 percent of the mass of Pluto, much the best match in the solar system. The gas giants' satellites include many small bits of rock, plus larger rocky and icy worlds, the pick of which are described later in the chapter. One, Titan, has a substantial atmosphere while Triton, a satellite of Neptune, is so large that it would be regarded as a planet in its own right if it were not in Neptune's grip.

Other objects

Finally, the solar system is home to several tens of thousands of asteroids, billions of comets, and uncounted numbers of dust particles and meteorites.

But are the planets of the solar system representative of the universe as a whole? To get the full picture, and gauge the Earth's uniqueness or otherwise as a planet, we need to find out about planets orbiting other stars.

Exoplanets

"Exoplanets" sounds cooler than "planets of stars other than the Sun", and is also less of a mouthful. Much creativity has been applied to detecting them. Initially, scientists attempted to detect slight movements of stars in the sky, which would indicate the influence of a planet's gravitation. But this has been tried for over sixty years with very limited success. More recently, astronomers have attempted to detect this movement indirectly, by measuring variations in the spectrum of the star's light reaching the Earth. As the star moves back and forth in the sky, its light gets a little redder or bluer, a change that can be detected with comparative ease. Most exoplanets have been found this way.

Another technique is to look at the "gravitational lensing" effect by which the mass of a planet will bend starlight. This has been increasingly productive. The "transit method" attempts to spot the dimming of a star as a planet passes in front of it as seen from Earth. And it is now becoming possible to look for the planets themselves, using powerful telescopes such as the Spitzer space telescope and the ground-based "Very Large Telescope" in Chile.

The problem is that the most successful of these methods, the search for changes in the spectrum of starlight, has a massive built-in bias. The planets it is most likely to detect are those which are heaviest, and nearest to the stars they orbit. That way you get the biggest gravitational pull on the star. Certainly any alien scientist using the technology now in use to detect exoplanets would not be able to spot the solar system by looking at the Sun. But in the next few years, this should begin to change. The lensing method, for example, is able to detect Earth-size planets. NASA plans a mission called Kepler which will probably detect them in abundance with this method. And a new European satellite called Corot should detect hundreds of exoplanets from 2007 onwards by watching for them as they transit the star they orbit.

At the moment, most known exoplanets are much bigger than Jupiter and have orbits far closer to their star than Mercury. This may tell us that our sample is too skewed to be worth worrying about. Such a solar system would be very different from our own, where the giant planets with all the gas are at a cool distance from the Sun while rocky ones with comparatively thin atmospheres, including Earth, are found closer in.

We are already obtaining knowledge of the atmospheres of exoplanets. One, for example, is so close to its star that it is streaming off hydrogen gas into space. It also contains sodium, carbon and oxygen – enough to get the biologists excited back on Earth.

Large telescopes now being planned – with names such as Owl, the **Overwhelmingly Large Telescope** – will be able to outgun existing telescopes in space and on the ground by some orders of magnitude. They should be able to get proper spectra of exoplanets and even produce simple images of them, as well as allowing us to decide whether the big planets we are detecting today are typical. At that point our ideas about the Earth's uniqueness will face a new test.

In the beginning...

So where did the Earth and all its fellow planets come from? In the beginning, the universe and all the matter in it essentially sprang from absolutely nothing, a minute "singularity" which erupted in a "**big bang**" to produce the expanding universe we see today. All this happened 14 billion years ago, and this was truly the very beginning: there is no point asking what there was before, because time itself, as we know it, was also generated at the Big Bang. This is because time slows down in gravitational fields, and that "singularity" contained all the matter in the universe.

The idea of a "big bang" seems counterintuitive. Indeed, the term was coined by the British cosmologist Sir Fred Hoyle (1915–2001) in an attempt to ridicule it. But in 1965 Arno Penzias and Robert Wilson found the leftover heat from the Bang – more formally known as the **cosmic background radiation** – which for most people settles the argument.

Since the Big Bang, the universe has been expanding steadily. We know this because of extremely elegant observations that show that further-away objects such as very distant galaxies are receding from us faster than nearby ones. This only matters on a cosmic scale and does not measurably affect the solar system or even a whole galaxy. If it seems further than it used to when you walk to the shops, you should take more exercise.

Forming galaxies

One thing that is important about the cosmic background radiation is that it is not absolutely even. In the jargon, it is **anisotropic**. This is another word of Greek origin and simply means that it looks different from different directions. The background radiation itself is incredibly faint, let alone the anisotropy, and it takes all our ingenuity to detect the very slight ripples that it contains. But these inhomogeneities reflect basic irregularities in the early universe, of which the background radiation is a sort of souvenir. If they had not been there, it is even possible that no matter would ever have clumped together to make solid objects. Instead, the atoms that make it up would just have got steadily further apart as the universe expanded.

As you are reading this, and sitting on a planet as you do so, we know that things did not work out this way. Instead, the slight irregularities in the universe allowed atom to meet atom and eventually entire galaxies to form. The structure that these irregularities produced is seen to this day when astronomers examine the distribution of galaxies in the sky. They are not randomly distributed but form clusters and other patterns. Some are very close, and collisions between galaxies are common. By contrast, stars within galaxies are comparatively far apart and close encounters between unrelated stars are rare.

A huge amount is now known about galaxies, including the fact that we live in one ourselves. The **Milky Way**, a band of light that crosses the night sky, had long been thought to consist of closely packed stars, and Galileo confirmed this by looking at it with his crude telescope. Now we know that when we look at the Milky Way we are looking at the disc of our own galaxy and that the centre of the galaxy is in the direction of the constellation Sagittarius. Here a huge **black hole** – matter collapsed into a miniature version of the singularity from which the universe itself sprang – forms the central mass around which the rest of the galaxy rotates. It takes the Sun about 225 million years to complete a single lap.

Making stars

Even within a galaxy, however, there is far too little material for everything to aggregate into stars or planets. Instead, matter has to be concentrated for the process to get going. You can see it in action if you get someone to show you the constellation Orion in the sky. As well as being big and distinctive, it lies across the Equator in the sky so it can be seen from New Zealand or Norway with equal ease in the southern hemisphere summer or the northern hemisphere winter. Just to the south of the three stars called Orion's Belt is a shiny, patchy bit. It consists of dense, glowing gas and dust. More precisely, it is dense by astronomy standards but is nowhere near as dense as the Earth's atmosphere which you are breathing now. It is opaque because it is as deep as a solar system. How it got that way is just being elucidated. The smart money backs the idea that a **shock wave** from an exploding star (or supernova) compresses material together until it is dense enough for stars to form. Similar pressure waves could also be produced when galaxies collide, and there is some evidence that this has happened in our corner of space. Within this mass of material, stars are being formed, so rapidly that it is almost possible to watch them turning on.

Many other such "**stellar nurseries**" have also been charted (see colour section p.1 for a photograph of one). In these areas, stars are forming in such dense and close-packed zones of space that most new stars exist as mutually orbiting double or multiple stars. However, since most older stars are so far apart, there must be some force that removes them from their siblings shortly after birth. Quite likely, gravitational disturbances in the gas cloud fling them out to make their way in the world.

Making atoms

The next part of our story demands a change of focus, from the vast dimensions of outer space to the atomic scale. The early universe produced by the Big Bang consisted mainly of one element, **hydrogen**. A hydrogen atom is the simplest atom possible, consisting of a proton with a positive charge being orbited by a negatively charged electron. These made up 74 percent of the universe while almost all the rest was **helium**, the next simplest atom, whose nucleus contains two protons and two neutrons, which as the name implies have no charge. A tiny amount of **lithium**, the third-lightest element, also formed in the Big Bang. The amount of helium in the universe is in agreement with theory and is powerful evidence that the Big Bang happened.

But look around and you do not see a world dominated by these three elements. Indeed, over 47 percent of the Earth's crust is **oxygen**, and **silicon** makes up another 28 percent. There is quite a bit of hydrogen about – they don't call water H_2O for nothing – but helium is so rare that it was identified on the Sun (hence the name) before it was found on Earth. Now it is extracted from natural gas (mainly in Amarillo, Texas, for some reason) to make balloons fly and for deep-sea divers to inhale. Lithium has some high-tech uses, such as the handy ability of lithium hydroxide to absorb carbon dioxide and prevent astronauts from dying of asphyxiation, but it is hardly central to life for the rest of us.

Instead, there is something else going on, and the big word for it is **nucleosynthesis**. The term simply means the synthesis (forming) of nuclei, the central cores of atoms. Everyone just assumes that once the nuclei are there, the electrons needed to orbit them will come, and this part of the theory seems to be sound.

The theory of nucleosynthesis is one of the great achievements of twentieth-century science, and the story it tells is a very satisfying and complete one.

Nucleosynthesis happens in two places. The creation of all that hydrogen, helium and lithium mentioned above is called **Big Bang nucleosynthesis** (BBN is the acronym if you need to impress astrophysicists). But the creation of all the other elements occurs in stars, in what is known as "**stellar nucleosynthesis**".

Normal stars like the Sun are powered by the energy released when four protons, or hydrogen nuclei, fuse to form one helium nucleus. This process, called "**hydrogen burning**", is a kind of nuclear fusion, and is the same force that powers a hydrogen bomb. But when two helium nuclei collide, there is no stable nucleus that can form. Only in stars older and hotter than the Sun (red giants and red supergiants) is there enough energy for "**helium burning**" (the fusion of three helium nuclei to produce **carbon**) to take place, and for even heavier elements such as **oxygen** to be formed, via these carbon nuclei accumulating other particles.

But even these stars cannot form the heaviest elements. Stars cannot gain energy by making elements heavier than **iron**, because it takes more energy to power the process than it produces. So these elements – such as uranium, plutonium, gold and silver – are only formed when the most massive stars explode as **supernovae**. At these extremely high temperatures neutrons are captured by existing nuclei to build up yet heavier atoms before being blasted into space by the exploding star. So any atom of gold you touch started life in the heart of a supernova. This super-stellar origin explains why elements that are important to us, such as **gold**, are vanishingly rare in the universe overall.

Making planets

So how do planets themselves form? A range of theories flourished in the nineteenth and twentieth centuries. In one, a passing star would have sucked a cigar-shaped mass of material from the Sun to cool in space and yield planets. This was a neat idea because it allowed the arrangement of the solar system's planets to be explained. Near the Sun, we find small, rocky planets (the Earth, plus Mercury, Venus and Mars). Go a little further out and you find bigger planets in the shape of Jupiter (biggest of the lot), Saturn, Uranus and Neptune. Pluto and other small objects lie beyond.

But the current view is a little simpler. As we have seen, as well as emitting shock waves which compress particles together to form stellar nurseries, exploding supernovae seed the surrounding area with heavy atoms. Through chemical reactions, these atoms build up into more complex molecules, which gradually stick together to form lumps called **planetesimals**. Over time, planets form from the collision and aggregation of these pieces.

A German-led group of scientists has run an experiment called CODAG, the Cosmic Dust Aggregation Experiment, on NASA's Space Shuttle to prove that dust particles in space can stick together in this way. To see the process in action on a larger scale, we can take a look at the star Beta Pictoris – astronomy code for the second brightest star in the constellation Pictor, the painter. Look at this star with a good enough telescope and you will see that it lies at the centre of a flat disc of muddy-looking dust. Similar dust discs have been observed around other stars and they confirm that when a star forms it does not absorb everything

The one that didn't make it

Not all the matter in the solar system has been swept into the Sun or the planets. There are many other objects out there, such as **comets**, **meteorites** and **asteroids**, although their combined mass is trivial, less than half the mass of the Moon. In the main, these do not concern us here; however, there is one part of their story which is relevant to this chapter's planetary interests.

We have known for a long time that meteorites found on Earth are either stony or iron-based. Some of the stony ones look like unaltered material from the early solar system. But others look more like terrestrial rocks. The only explanation for how they got this way is that they must have been built into a planet that later broke up. Similarly, the iron ones contain patterns that show they must have cooled very slowly, probably in the core of a planet. There are even a few "stony irons" that seem to have come from the boundary between the iron core and the rocky outer layer of a planet.

This long-gone planet from the early solar system must have been tiny compared to the Earth. To judge from the leftover parts that arrive on the Earth today, it never developed oceans and weather of its own, much less life. The rocks are quite unlike the sedimentary types formed by water and weather at the surface of the Earth.

around it. Instead, at least some stars form at the centre of a spinning disc of material. Most ends up in the star but enough is left over to produce planets. Some never does get captured by the star or a planet and remains to yield comets and other small objects.

In a solar system like this one, the star at the centre contains most of the mass, but consists almost entirely of helium and hydrogen. As the star forms and begins to shine, pressure from the light it emits stops more dust and gas falling in. This means that heavier elements are concentrated further out.

How the Moon was made

There is one satellite that does not need me to tell you who discovered it, because at an early age, you discovered it for yourself. The Moon has been recorded – to begin with, in carvings on animal bone – for over 20,000 years. It affects our lives in too many ways to describe, and in the next chapter we will see how it forms an active part of the Earth system.

About 16 percent of the Moon's surface is covered in big, flat round areas. They are so featureless, especially to the small telescopes of early astronomers, that they are called **maria**, or seas, although we now know that no ship has ever sailed on them. Instead, they are solid layers of lava produced at the last stage of the solar system's formation when the Moon was bombarded by large planetesimals. Their impact produced so much heat that the immense craters they generated filled with molten rock. They are smooth today because they solidified that way and have only a thin splatter of later craters caused by meteorite impact. Many of the features around them also show signs of lava flooding.

By contrast, the rest of the Moon is made up of the **highlands**, and if space tourism ever gets going, this will be the destination of choice for the honeymooners. The reason is that the highlands have survived from the earliest days of the solid Moon. They are heavily cratered, and in many cases it is possible to see new craters that have obliterated parts of old ones beneath.

In the last few decades, we have found out more about the Moon for several reasons. One is that between 1969 and 1972, a dozen Americans visited the place, had a look, and brought bits back. In addition, a clutch of US, Soviet and European spacecraft have visited the Moon, and the Soviets have also returned Moon rock to the Earth. Even more startlingly, it turns out that some of the meteorites found lying about on the Earth have come from the Moon.

When scientists studied the atomic make-up of Moon rock, they were in for a surprise. Apart from some subtle differences, its composition is very like that of Earth rock. In recent years, scientists have converged on the reason why, and in the process they have found out how the Moon formed.

Near a star's fierce heat, planets struggle to hold on to water and gases and tend to be almost entirely rock. The nearest planet to the Sun, Mercury, orbits at about 58 million km out, compared to 150 million km for the Earth, and its only atmosphere is a feeble array of passing atoms emitted from the Sun itself. Further out, Venus, Earth and Mars (known as the terrestrial planets) all have reasonably useful atmospheres, but they do not add up to much compared with those of the big planets of the outer solar system, which are often – and with reason – termed the gas giants. Their cores are not unlike those of the inner rocky planets, but they are

It is simple to form an object as big as the Moon – there are about fifteen planets and satellites in the solar system of about its size or bigger. But it is more tricky to develop a computer model of the early solar system in which the Moon can form very close to the Earth without their aggregating into one. And models which involve the Earth capturing the Moon after it formed tend to have annoyingly artificial assumptions, raising more problems than they solve. However, the data about the composition of Moon rock suggests that the Moon is in fact a visible relic of the violent formation of the solar system. There is now general agreement that the Moon was formed when a chunk of matter the size of Mars hit the Earth a glancing blow, blasting a massive piece away, which later settled down to form the civilized sphere we see today. Because the Earth's surface changes so fast, the immense crater that this impact must have left is long gone. It would have topped this book's list of must-see Earth sights by some distance.

The lunar crater Daedalus photographed by Apollo 11 astronauts in 1969. Terraces of slumped material are visible in the walls of this 93km impact structure.

overlain by many thousands of kilometres of gas which has its own chemistry and weather.

How long does all this planet-making take? Not long by geological standards. In the outer solar system, Neptune and Uranus, despite being far bigger than the Earth, formed in about 10 million years. Nearer the Sun, it takes a little longer. The Earth took about 100 million years to form. Its core would have taken 29 million years to accumulate, as against 13 million for Mars's much smaller core. But the biggest object in the solar system apart from the Sun itself, Jupiter, will have formed very rapidly, with its core building up in only 100,000 years.

The Earth beneath our feet

Now that we've charted its origins, let's take a look at planet Earth in a little more detail. The centre of the Earth is a chunk of hard metal, mostly iron, called the **inner core**. It has a radius of about 1200km; as is often pointed out, that is about the size of the Moon. Above that, you find more iron, but this time molten. This **outer core** is about 2200km thick. It is heated by warmth mainly left over from the gravitational energy emitted as the planetesimals fell together to form the Earth. The movement of this molten metal is the generator that creates the Earth's magnetic field (see p.121).

Move a little further out and you start to arrive at material looking more like "rock" as we know it. This is called the **mantle** and is about 2900km thick.

Although the mantle consists mainly of material that we would recognize as rock, it is so hot and under so much pressure that it too is liquid, or at least viscous. Its flow provides the motive power for the movement of the continents, as the upper mantle takes the topmost part of the solid Earth, the **crust**, with it as it moves. The crust is on average 35km thick.

Finally, the solid Earth is overlain with yet more fluids. The **atmosphere** and the oceans and seas (the **hydrosphere**, for Greek enthusiasts) are the most obvious, but don't forget the **cryosphere** (the frozen bits, ice and snow) and the **magnetosphere**, the Earth's magnetic field and the charged atomic particles that it captures, mainly from the Sun. And of course there is the Earth's gravitational field, which continues with ever-diminishing force across the whole of the universe.

How old is the Earth?

There is no direct way to find out the age of the Earth. In the modern era, the predictably steady decay of radioactive material in rocks allows them to be dated. But even that is little help here. The Earth is so geologically active that it has no rocks lying around from the time of its formation. Instead, the Earth has to be dated more by analogy with meteorites and other objects thought to have formed at around the same time. This yields an age for the Earth of about 4.54 billion years.

But in the past, the lack of hard data has not deterred scientists from coming up with their own estimates, based on some ingenious lateral thinking.

For example, it is obvious that the Earth and the Sun formed by condensing from some much larger and less dense mass. Lord **Kelvin**, one of the greatest scientists of the nineteenth century, was quick to see that this condensation must release energy in the form of heat. And as befits the man who gave his name to the scientific scale of temperature (hence the K), Kelvin saw that this was the high road to determining the age of both objects. He worked out that for the Sun to be as hot as it is today, it could not be more than a few million years old. He then applied the same logic to the Earth, using figures for the heat flow through the crust, and worked out that the Earth could not be more than 40 million years old.

These figures were published to the distress of a Victorian England just getting used to the idea of evolution by natural selection made famous by **Charles Darwin**. Darwin thought that the Earth had to be ten times as old as Kelvin had "proved".

As the saying goes, idiots do pretty stupid things, but to make a terrible mess you need someone really clever. A few years after Kelvin's figures were published, radioactivity was discovered and it turned out to be the key to the problem. Part of the Earth's inner heat comes from the decay of uranium and thorium atoms (mainly), while the Sun's comes from fusing hydrogen into helium.

Before you scoff, think of it from Kelvin's point of view. At that time, even the idea of atoms was by no means universally accepted, much less the concept of them decaying to make the Earth warm inside and fusing to make the stars shine. And in any case, his estimate was far closer than the few thousand years favoured by earlier scholars such as the Bishop of Armagh in Ireland, **James Ussher**, who calculated that the Creation had occurred on the evening of 23 October 4004 BC. Kelvin attempted to apply science to the problem and was out by a factor of 100, but Ussher's answer was 750,000 times too short.

However, Kelvin will go down in history as a lesson that distinguished scientists can be dead wrong. He announced in August 1896 that he had "not the smallest molecule of faith in aeronautical navigation". The Wright Brothers took to the air seven years later.

Venus, the Earth's unlucky twin

Venus is 12,100km across, 95 percent the diameter of the Earth. It has 82 percent of its mass, 90 percent of its surface area, 90 percent of its surface gravity... You get the idea. It even seems to have a core, mantle and crust much like the Earth's, although its feeble magnetism rules out a molten iron core.

But this is where the similarities end. For one thing, Venus takes 243 days to rotate on its axis, compared to one day for the Earth, and does it backwards compared to us. The Earth is rotating at over 1600kph at the Equator, but on Venus the figure is 6kph.

And then there is that **atmosphere**. Venus is the brightest object in the sky apart from the Sun and the Moon. Part of the reason is that it is near – it comes nearer than any other planet to the Earth. But the other reason is that its albedo (see p.21) is massively high at 65 percent. That is because it is covered with continuous cloud, and we are seeing the silver lining.

Despite being further from the Sun than Mercury, Venus is far hotter, with surface temperatures exceeding 500°C. The reason is that Venus has the severest case of the **greenhouse effect** (see p.250) in the known universe. The dense clouds trap heat in abundance – indeed they also transmit it so effectively around the planet that the night-time side is pretty much as hot as the sunlit hemisphere.

The Earth's greenhouse effect is mainly caused by **carbon dioxide**, but this makes up less than 0.1 percent of the Earth's atmosphere. It accounts for 97 percent of the Venusian atmosphere, with nitrogen making up most of the rest. Some global warming activists regard Venus as an awful warning about our own future if we burn too much fossil fuel. But in fact there is no way things on Earth can work out remotely as they have on Venus.

Venus's dense atmosphere means that – with the honourable exception of fewer than a hundred pictures taken by the Soviet Union's Venera landers – nobody has ever seen its surface. But it has nevertheless been mapped in some detail. Most of the hard work was done by NASA's Magellan spacecraft, which lugged a 3.4 tonne radar to Venus to map the surface from orbit. As a result, all the big features have been identified and named – after women, an inspired decision by the International Astronomical Union.

Magellan showed a surface with few of the small meteor craters that litter the Moon or Mars. The atmosphere is too thick for any but the biggest meteorites to get through. It is also so dense that it erodes the surface almost as water does on the Earth. The big difference between Venus and the Earth, however, is that Venus is without active surface-building like the plate tectonics that constantly redraws the map of the Earth. The reason is probably that the crust is too stiff and thick to be mobile. Although there is a liquid mantle beneath, it cannot shift the crusty carapace on top. Instead, it seems to have catastrophic volcanic eruptions every so often, most recently about 800 million years ago, that completely remake the surface with fresh basalt lava.

Earth cousins

How unique is the Earth in its structure and composition? Our knowledge of the other planets in our solar system has been increasing apace in recent times. As the astronomer Carl Sagan (1934–96) put it, we are the generation in whose lifetime the planets of the solar system have turned from being ill-seen lights in the sky to being familiar worlds. We now have the surface of Mars essentially under continuous observation. We know that it has had **rivers** – perhaps recently – and live **volcanoes**. And we know that Mercury has a **magnetic field** not too unlike the Earth's, suggesting it probably has a big **iron core** like the Earth's, or did comparatively recently. So far Mercury has only been visited by one spacecraft, *Mariner 10*, although at the time of writing the US Messenger spacecraft was on the way, while Europe's Bepi-Colombo Mercury mission was at the planning stage.

It might seem unlikely that the outer solar system contains planets with much in common with the Earth. But although the gas giants seem radically different from the Earth, some of their satellites do resemble our own planet in certain ways. One of Jupiter's satellites, Io, has what may be the only active volcanoes in the solar system beyond the Earth. These volcanoes have lava flows of molten rock like those found on the Earth. The plumes of material they emit into space are mainly sulphur dioxide – itself a common component of volcanic emissions on the Earth. However, while Etna and Mount St Helens are powered by heat generated deep inside the Earth when radioactive atoms disintegrate, Io's volcanoes get their energy from the gravitational energy of Jupiter itself.

Two of these satellites in particular merit a closer look.

Titan

Titan is a satellite of Saturn discovered by the Dutch astronomer Christiaan Huygens in 1655. It is smaller than the Earth, about 5150km across to the Earth's 12,800km. Given everything we know about the formation of the solar system, it is almost certainly about the same age (4.54 billion years) as the Earth and the other major planets, having been formed by aggregation at the same time. It is certainly a rocky world like the Earth because its density is very similar to the Earth's.

Titan is one of the few places we know (another is Venus, see box) to share a key attribute of the Earth – a thick, dense **atmosphere**. In fact, like Venus, Titan's atmosphere is so dense that passing spacecraft, or telescopes on Earth, can never see the surface in visible light.

Titan even has a **weather system** as the Earth does. In 2005 a European spacecraft called *Huygens* was launched into Titan's atmosphere to take a look. It found a world which seemed eerie in its familiarity to human onlookers. The atmosphere had rain, wind and clouds. On the surface could be seen seas, rivers, lakes, bays and valleys and, at the smallest scale, rocks.

However, all this is cosmic trickery on a grand scale. Like the Earth's atmosphere, Titan's is mostly nitrogen, a gas known for its reluctance to get involved in any serious chemistry, at least by comparison with oxygen. But instead of oxygen, which makes up most of the rest of the Earth's atmosphere, the remainder of Titan's atmosphere is made up of **methane**, the smallest possible molecule of carbon and hydrogen, and its close chemical relatives, collectively called the hydrocarbons. The rocks lying about are in fact the only thing there made of any familiar material, and even they seem to be mainly lumps of ice rather than stone. The seas and rivers are mostly of methane and so is the rain. To us, methane is the "natural gas" that we use for cooking, although it is also shipped around the world in liquid form in refrigerated containers. On Earth, methane mainly comes from rotting vegetable matter trapped in the ground, as do coal and oil, but on Titan it was probably formed by chemical processes in the early solar system.

It is certainly possible to observe something very like the **erosion** that goes on at the Earth's surface in action on Titan, with the difference that on Earth, water does most of the hard work. On Titan, deposits of dark hydrocarbons are washed away by methane rain.

But although erosion and the other processes at work there will have changed the detailed map of Titan over time, Huygens would have seen essentially the same phenomena at any time in the last few billion years. By contrast the Earth is a dynamic planet with active systems that change its surface the whole time. With the arrival of human beings, it has become even more exciting, with big changes happening fast. Io is probably the only quicker-changing object in the solar system. Its active volcanoes rework the surface so fast that there is no point in taking a map if you are sent there as an astronaut. It will be out of date before any possible spacecraft can get you there.

There is one further difference between the Earth and Titan: while the Earth is about 150 million km from the Sun, Titan – like Saturn and the rest of its extensive satellite entourage – is over 1.4 billion km away, nearly ten times as far. This means that unlike the Earth, which basks in a non-stop flow of solar energy, the surface temperature on Titan is minus 180°C.

This is responsible for the most significant difference of all between Earth and Titan: while the Earth has a rich and varied biology, Titan is devoid of life. The Earth's proximity to the Sun has allowed **photosynthesis** (the process by which plants use light to grow and release oxygen) to take place, transforming the planet's surface.

Europa: the real Waterworld

Another of Jupiter's satellites, Europa, was discovered by Galileo in 1610 along with three others. In the pre-space-probe era, Europa fascinated scientists by being one of the brightest objects in the solar system, reflecting back 64 percent of the light that fell on it. (This figure is called the **albedo**; in comparison, the Earth's is about 30 percent.)

Once we got a closer view, Europa, like Titan, turned out to look oddly familiar. But unlike Titan, Europa has almost no atmosphere. No thick clouds defend the surface from the intense cold of outer space. What slight atmosphere Europa does have consists almost entirely of oxygen. The reason is that the planet's surface is covered with **ice**, and cosmic rays striking it split the water into hydrogen (which escapes) and oxygen, which is held back because its atoms are heavier.

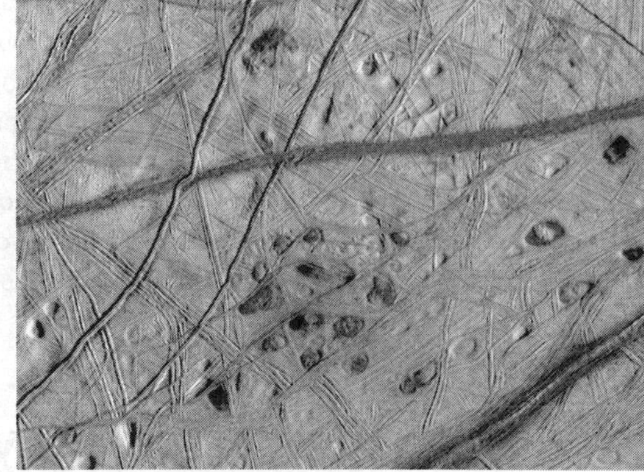

The cracked ice floes of Europa are underlain by liquid water. The coloured spots are probably points at which water from this ocean has reached the surface, and may hold clues to its composition.

The surface looks like nothing so much as an aerial photograph of the Arctic Ocean, complete with ice floes crunching together and throwing up raised icebergs. Indeed, if you believe current wisdom about global warming on Earth, Europa may have some educational value in a few years' time: it may become the last place in the solar system to feature pack ice. There are even remains of bizarre geysers or volcanoes, caused by the movement of the ice. When it shifts and a gap opens up to the water beneath, huge amounts of water vapour spray out into the vacuum of space until enough of it has built up to close the rift.

Before the space age, our telescope-based knowledge of the solar system told us mainly about its inner planets, with the result that **water** was thought to be rare except on Earth. We now know that water is common in the outer solar system. For example, another of Saturn's moons, **Enceladus**, has an even higher albedo than Europa (about 75 percent) and once again its surface is pretty much solid ice. But Europa seems to be a unique ally of the Earth, at least in the known universe, in having a supply not only of ice but of liquid water. At more than five times the Earth's distance from the Sun, how does the water avoid freezing? The answer is probably that, as with Io, Jupiter's gravity sets up forces within Europa that turn into heat. The result is a deep ocean with a thin crust of ice – perhaps about 5km thick – formed by the cooling effect of deep space.

Europa may have lessons for our thinking about the **origin of life** here on Earth. Somewhere that has abundant liquid water and heat, after all, has in place several of the assets that make it possible for life to exist on Earth. There are even proposals to send a space probe to Europa which could melt a hole in the ice and swim about beneath to look for any passing haddock. While they are there, they would be wise to head on down to the ocean floor. As well as the water volcanoes on the surface, there may be more normal volcanism going on at the point where the water meets Europa's rocky core. On Earth such locations are a prime site for life, and maybe the same is true of Europa. This would, however, be some trip for the robot concerned as the ocean is probably about 20km deep, and it will be at least a decade before this question is answered.

Is the Earth the only planet to harbour life?

Pending the results of any expedition beneath Europa's icy exterior, and of investigations into the possibility of life on Mars (see box overleaf), the Earth is currently the only planet that we know houses life. This seems obvious if you glance about the solar system, where everything you see is too hot, too cold, too poisonous or in some other way too unwelcoming to let life get going or continue.

But think about it a little more and there seems to be something odd going on here. In the sixteenth and seventeenth centuries, it was finally proved that the Earth is just another planet, not the centre of the universe. Later it became apparent that despite being unique in many ways, humans are just another species, produced by natural selection like all the others. Now it is generally accepted that even the universe is probably one of many.

Can we really be sure that there is no other life in the universe? Of course we cannot. But in the 1960s the British scientist **James Lovelock** devised a neat way of detecting life on any planet with an atmosphere whose composition can be determined. (Lovelock appears later in this book as the founder of Gaia theory, see p.256.) He reasoned that if the atmosphere contains two gases that would normally react together and deplete one another, there must be some living creature on the planet to go on producing one or both of them. The obvious example is the oxygen in the Earth's atmosphere. Oxygen is highly reactive, which is why firefighters deserve to be well paid. It exists in the atmosphere alongside methane, even though the two react together. The only reason they co-exist is that the oxygen is constantly replenished by fresh supplies generated by plants. The methane comes partly from living organisms and partly from inorganic sources such as volcanoes.

This argument is not completely foolproof. It is possible to imagine a world in which two conflicting atmospheric gases are generated in different places without the involvement of life and fight it out in the atmosphere. But we are now getting to the stage at which it is becoming possible to measure the atmospheres of planets of other stars. Maybe, as Lovelock foresaw, Earth-like incompatibilities will emerge that argue for life on these planets.

The real issue, perhaps, is that our thinking about life on other planets is determined by what we know here. After all, why do science fiction writers usually put their aliens on planets in the first place? And solid, rocky ones at that? Obviously this makes life simple for the human protagonists, but is there any reason why deep clouds such as those of Jupiter, or a gas cloud in deep space, could not be a handy spot for life? Thinking about life in the universe is a little like thinking about languages if the only one you speak is English. Even one more example would expand our wisdom enormously.

With this in mind, there are a number of approaches now being mined to find out whether there is any detectable life out there. The title of this endeavour is SETI – the **Search for Extraterrestrial Intelligence** – which gives the game away. What is being sought is not life, but signals generated by intelligent life not too unlike us. At the moment the emphasis is on trawling radio signals, but there are also proposals to detect signals of extraterrestrial life in visible light or even atomic particles.

The real question, however, is that posed by the Italian physicist Enrico Fermi and now generally called the **Fermi Paradox**. If the aliens exist,

Mars as the abode of life

Mars As The Abode Of Life was the title of a book published in 1908 by Percival Lowell. He was convinced that the canals of Mars were proof positive of the existence of an advanced civilization that had built a planet-wide irrigation system to ship water from the polar ice-caps to the Equator. At the time, the idea was in the air. A decade earlier, English science-fiction writer H.G. Wells had written a fictionalized account of a Martian invasion of the Earth, *The War Of The Worlds*.

In fact, the canals of Mars are one of the biggest examples in history of the danger of sloppy translation. These apparent lines on the Martian surface were first observed by the Italian astronomer Giovanni Schiaparelli in 1877. He called them **canali**, which in Italian means channels, and he never meant there to be any implication that they were artificial. Only when the word got mistranslated into English as canals did that idea get going.

It has now been demonstrated that the canali were nothing but optical illusions. But the detailed knowledge we now have of the Martian surface suggests that although the locals have not carried out any major civil engineering, it is impossible to rule out the possibility of Martian life on a modest scale in the past, and it may even cling on today.

All views of the Martian surface show something that on Earth would be regarded as conclusive proof of the presence of **water**. On a large scale, there are features resembling river valleys, while small-scale views show layered rocks and apparent dried-out mud and pebbles that to the eyes of Earthly geologists can only have been created by water. They may be just as wrong as Lowell, but they are working from a stronger evidence base.

Because water is regarded by humans as a definite must for life, this discovery encourages belief in Martians, albeit small ones. In addition, there are signs that some of the water-sculpted landscapes of Mars may be recent.

But is there any more direct proof of life on Mars? As long ago as 1976, two US Viking spacecraft landed on Mars and put some Martian soil through an experiment designed to detect biological activity. The argument about the results is still going on, with some scientists believing they do show life but NASA being far more cautious.

Another approach is to look at the Mars **meteorites** that sometimes land on the Earth. Thirteen of them are known. One landed in 1911 at a place called Nakhla where are they? They have had plenty of time to get here, after all, if you compare the speed of a possible spacecraft to the size of the galaxy. But despite the loose talk about UFOs, there are hundreds of thousands of skilled amateur astronomers looking at the sky every night and they never see anything they cannot identify.

in Egypt and killed a dog. But the one called Allen Hills 84001, found in Antarctica in 1984, is more interesting because it contains tiny spherules that may be **fossil bacteria**. Or they may be inorganic crystals. Or they may be Earthly bacteria that have found their way into the meteorite while it lay on the Earth's surface. The meteorite also contains minerals that may have been deposited by bacteria, and perhaps organic compounds. Either way, the rock left Mars billions of years ago and the species it contains, if any, may since have become extinct.

In 2004, **methane** was detected in the atmosphere of Mars by telescopes on Earth and by a European satellite orbiting Mars. Methane gets broken down in the Martian atmosphere, so if it is there, it is being replenished by some means. On Earth there are several sources of methane but nobody can find any very obvious ones on Mars. Mars has plenty of dead volcanoes, a prime source on Earth, but no live ones have been glimpsed. Another obvious methane source is bacteria, which would be consistent with the pro-life view of the Viking results.

The new generation of Mars spacecraft promises to sort all this out in the near future. If the answer is a yes, the headlines will be unmissable.

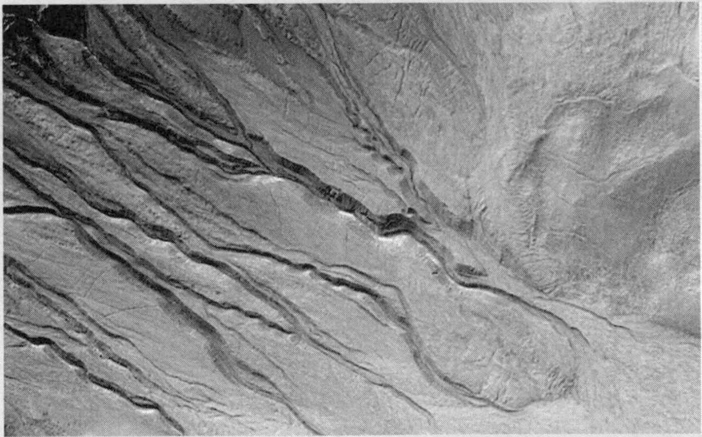

These gullies on the surface of Mars look fresh, as the dark material in them has not been covered over by wind-blown dust. It suggests that the area has seen liquid water in recent years.

A number of answers have been thought up for this question. One is that there are indeed no aliens. Another is that because of the crude state of human development, the Earth has been set aside as a kind of wildlife park which aliens have agreed not to visit. But perhaps this explanation also relies too much on the assumption that other species are like us, with a restless urge to explore new worlds. Maybe they are happy just where they are.

2

The Earth in space

The Earth in space

As we saw in Chapter 1, a planet is almost by definition a subsidiary body. So when thinking about the Earth, it makes sense to start by thinking of it in its setting, in the context of the forces that surround it. That means everything from the Moon and passing meteorites to the star that keeps the whole thing going – the Sun.

Like every star, the Sun is really a pretty simple device – just a very hot centre where heat is being generated by nuclear fusion, surrounded by hot gas through which it is transmitted to the universe beyond. Astrophysicists can describe the whole thing in some detail in a few equations. As we shall see, there are magnetic and electrical effects going on that complicate matters and have severe effects here on Earth. But a complete description of a square kilometre of the Earth's surface would be far longer than a detailed account of the workings of the entire Sun. The Sun's effects on the Earth are, however, far from simple.

The seasons

From our point of view, the Sun's main purpose is to keep us warm. The Earth receives energy from the Sun at a rate of about 340 watts for every square metre of its surface, mostly in the form of heat or light. Of this energy, 30 percent is reflected away at once, and the rest is eventually radiated back by the sea, the land and clouds. If the amount leaving did not match the amount arriving, the Earth would either cool down or heat up.

The warmth the Sun provides is not equally distributed around the globe. Instead, we have the phenomenon of **the seasons**. They arise because the Earth's orbit around the Sun is at an angle – about 23.5° – to

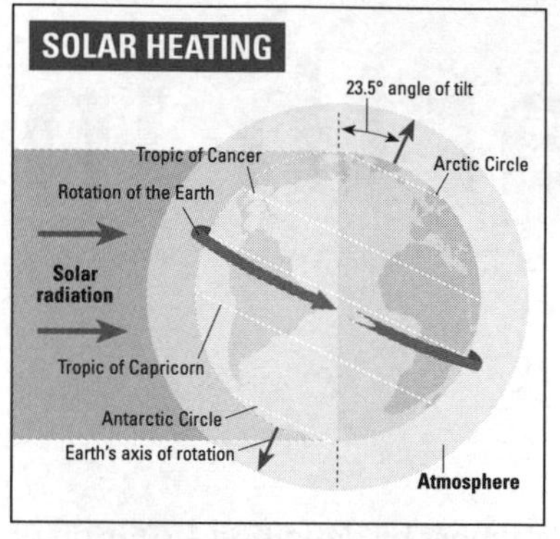

the Equator. (That is why the Equator on a globe of the Earth is tilted at this angle to the horizontal.)

The result is that in addition to moving across the sky from east to west each day as the Earth spins on its axis, the Sun moves from north to south in the sky over the course of a year, northwards during the northern summer, and south during the northern winter. This explains why summer is hotter than winter. If the Sun is higher in the sky, its heat is shining down on the Earth below more steeply. The effect is like shining a torch at a wall. If you shine it straight at the wall, the light is brighter than if you shine it at an angle because the same amount of energy is hitting a smaller area. In addition, the nearer the Sun is to your latitude on Earth, the more time it spends above the horizon and the longer the day becomes. (See www.cru.uea.ac.uk/~timo/sunclock.htm – a handy website that shows a sunclock, giving the Sun's position above the Earth in real time. It's much more fun than it sounds.)

Things are further complicated by the fact that the Earth's orbit around the Sun is not an exact circle. Like all planetary orbits – as **Johannes**

Dividing up the globe

Four imaginary lines help describe how the Sun's position in the sky varies according to your latitude. The **Tropic of Cancer** and the **Tropic of Capricorn** are lines – 23.5° north and south of the Equator – between which the Sun will be vertically overhead at some point in the year. They get their names because the Sun is in the constellation Cancer when it is at its furthest north (in June), and in Capricorn when it is at its extreme south (in December). The term "the tropics" can also be used to refer to the area of the Earth that lies between the two.

At the other extreme are the **Arctic and Antarctic Circles**, 23.5° from the poles. They are the lines beyond which the Sun does not get above the horizon for some part of the year. The nearer you get to the poles, the more extreme things become. At the poles themselves, the Sun rises above the horizon as it crosses the Equator and stays there for six months until it is back at the Equator again. Then six months of darkness ensue, which is much less enjoyable.

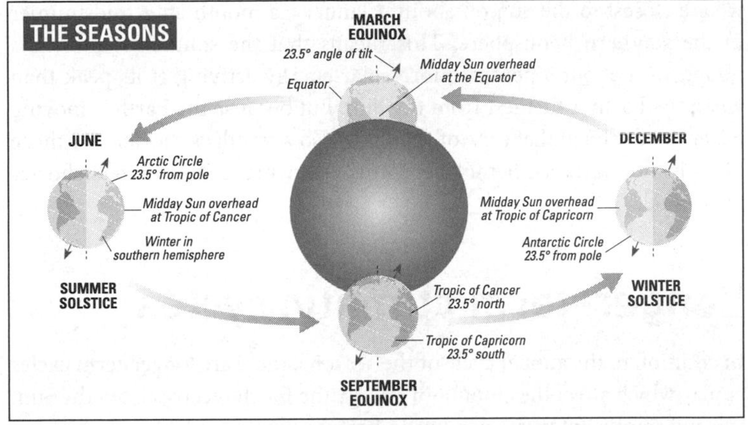

Kepler pointed out in 1609 – it is an ellipse, differing from exact circularity by about 1.7 percent. As Kepler also showed, the nearer a planet is to the Sun, the faster it moves. (The same applies to other objects in the solar system: comets that dawdle along at a snail's pace for thousands of years beyond the orbit of Pluto move about 100 times as fast as Concorde at full tilt when they arrive in the inner solar system.) In the Earth's case,

Kepler's laws of planetary motion

Johannes Kepler set out his first and second laws of planetary motion in 1609 and the third in 1618.

The first states that **the orbit of a planet is an ellipse with the Sun at one of its foci.** The foci of an ellipse are hard to define but easy to show with a nursery experiment. Push two nails into a surface maybe 20cm apart, throw a loop of string round them, put a pencil in the loop and run it round. The shape you'll have drawn is an ellipse and the nails are the foci. The Earth's orbit is nearly circular, but its slight eccentricity – the difference between its shape and an exact circle – has tangible effects, as the text explains.

The second law states that a line from a planet to the Sun sweeps out equal areas in equal intervals of time. In other words, **the closer a planet gets to the Sun the faster it goes**. Again, this has real effects for the Earth despite the low eccentricity of the Earth's orbit.

The third law states that the period a planet takes to orbit the Sun squared is proportional to the size of its orbit (to be exact, the long radius of the ellipse) cubed. In other words, **planets deeper in the solar system move much slower**.

we are closest to the Sun on about 4 January, a month after midsummer in the southern hemisphere. This means that the southern summer is hotter, with about 3 percent more solar energy arriving at its peak than when the Earth is furthest from the Sun. But because the Earth is moving faster in its orbit at that time of year, it is also a few days shorter. For those of us in the northern hemisphere, this effect makes the winters shorter and milder.

Longer-term climatic cycles

In addition to the annual cycle of the seasons, there are longer-term cycles at play which affect the amount of energy the Earth receives from the Sun, causing significant variations in the Earth's climate.

First, there have now been almost 400 years of routine observations of the Sun which show that **sunspots** appear on its surface with a 22-year periodicity. Sunspots are areas of the sun marked by intense magnetic activity and a slightly lower temperature than surrounding regions. They look dark only because we see them against the Sun's surface. In fact they are almost as bright (and hot) as the rest of the Sun. If one were magically transplanted to the night sky it would be dazzlingly bright. They are the signs that the Sun is especially active and is pushing out more energy than usual into space (see box overleaf). As we shall see in Chapter 8, there are those who claim that these changes in the Sun's power output are the secret behind global warming and that it has nothing to do with human activity. (To show my own bias, I also note that these tend to be the same people who claim that no warming is going on anyway, but we'll come to that later.)

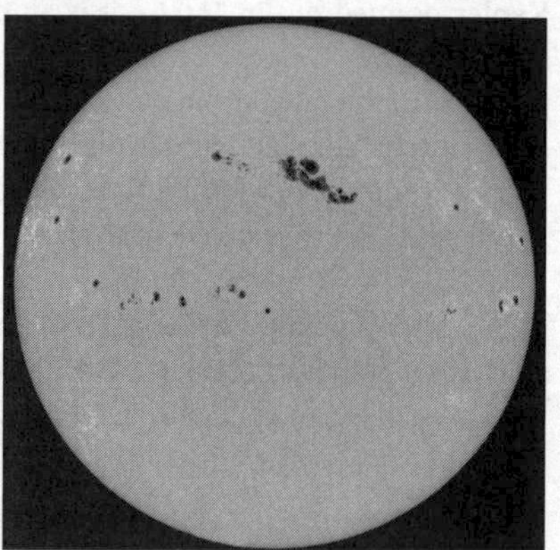

Sunspots. **See colour section p.1 for a close-up view.**

On a longer timescale are the "**Milankovich cycles**", named after Serbian astronomer Milutin Milankovich. First, the 23° **tilt** of the Earth's

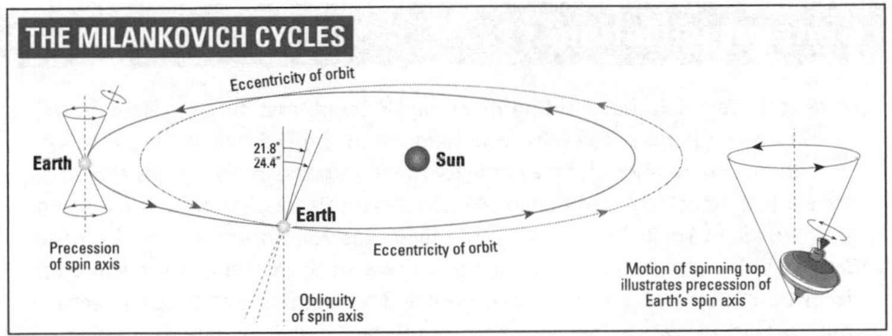

orbit to the Equator is not fixed, but varies between about 22° and 25° over a period of around 41,000 years. When the tilt is greater, the seasons get more severe, with hotter summers and colder winters. The latter is likely to be the more important of the two. It means that more snow will fall and stick, and so increases the chances of an ice age taking hold (see p.219).

Secondly, there is the **precession of the equinoxes**, discovered by the astronomer Hipparchus in the second century BC. Imagine the Earth spinning like a top. Then imagine the way a top which is slowing down tilts over, so that the direction of its spin keeps moving around in a circle. That is precession. It takes about 22,000 years for a full loop to be completed.

This precession has a number of effects. One is that maps of the sky get out of date, so you need to specify the year in which they were accurate – for example, the Pole Star that helps people in the northern hemisphere to navigate has not always been near the North Pole. More importantly for our purposes, it means that 11,000 years ago the Earth was at its closest to the Sun in July instead of January, northern hemisphere summer rather than southern. Again, this might encourage ice ages by allowing more snow and ice to form in the northern hemisphere, where most of the big land masses are (glance at a map or globe to confirm this).

Finally in this cosmic dance, even the Earth's orbit is not quite the constant you might think. Its **eccentricity** varies over two cycles of 100,000 and 400,000 years.

Milankovich claimed in 1920 that these changes were the cause of ice ages, and this view now has strong support. More recently, the late US astronomer Carl Sagan and others pointed out that the axial tilt of Mars might vary far more than the Earth's, which is kept in check by the gravitation of the Moon. So Mars could have more extreme variations in its climate than the Earth and could have been far warmer and more welcoming a few tens of thousands of years ago.

Sunspot variation

Have you ever seen those old prints of happy Londoners holding "**frost fairs**" on the frozen Thames? The fairs were held when the Thames flowing through London froze hard enough for people to move onto the ice for weeks at a time. The earliest seems to have been in 1434. In the 1660s, ice skating was in fashion on the river. In 1683–84, the ice on the river was 28cm thick (known in those days as 11 inches). And in 1715–16, the ice was so thick that a storm surge up the river lifted it by 5m but it did not break. The first fair with properly organized socializing was in 1564–65, and there was an epic as late as 1814 with eating and donkey rides on the ice, and whole streets of shops appearing. For more information see:

H2G2 Frost Fairs Guide bbc.co.uk/dna/h2g2/A970733

A frost fair on the Thames in 1683–84

The tides

While the Sun gives out the heat and light that makes life on Earth possible, it also contains almost all the mass in the solar system, which means that it has almost all of the gravitational pull. This keeps the Earth in its orbit, but that is not the whole story. Set loose on the rotating Earth, the Sun's gravitation has more subtle effects. More to the point, so does the gravitation of the Moon, which is more powerful because although the Moon is far smaller, the Sun is four hundred times as far away.

Well, I have lived in London since 1973 and have yet to spot pack ice. There seem to be two reasons why the fairs became possible and why they died out. The first is old London Bridge, removed and replaced in the 1830s. It was a massive structure with shops and houses, and ponded water behind it, slowing the flow of the river. The new bridge is much sleeker and lets the water flow more freely so it is less likely to freeze. However, the main reason the river no longer freezes is that things were a lot cooler back then. Indeed, it was not just the Thames that froze. There are accounts of people walking from Staten Island to Manhattan in 1780.

Part of the reason seems to be that the Sun was a lot less lively in that period. From about 1400 to 1800, it seems to have had fewer spots on its surface than we see today.

This view is problematic, since systematic telescope observation of the Sun only dates from the seventeenth century. For information before that period we are dependent upon patchy, naked-eye observations. Between 1645 and 1715, a period known as the **Maunder Minimum**, sunspot activity seems to have been at its lowest, but at this time the practice of sunspot-spotting was in its infancy.

However, sunspot dearths later in the period, such as the **Dalton Minimum**, from 1790 to 1820, are more solidly established. And it is certainly true that temperatures were significantly cooler at this time, earning it the name the "**Little Ice Age**".

The Little Ice Age was preceded by the "**Medieval Warm Period**", which lasted from around 1000 to 1300. The end of the Medieval Warm Period seems to have been a factor in the disappearance of the Viking colony in Greenland, which vanished from history at this time. It was also responsible for relegating England from the first division of wine-making nations to the lower leagues, closing down most of the vineyards of southern England – although the current period of global warming may see more vines returning to England's slopes.

Acting on the Earth, the gravitation of these two bodies produces effects known as **tides**. You probably think of tides – rightly – as the thing that makes the sea go up and down the beach when you are on holiday. Only if you sail – anything from a dinghy to an aircraft carrier – do you need to pay close attention. However, the tides in the oceans are only part of the story. Lunar and solar gravitation also raise detectable tides in the atmosphere, and in the solid Earth itself.

Air tides seem to have been detected for the first time in 1918. There have been suggestions that they are related to extreme weather. But these claims are disputed, since the effects of air tides are very small, far smaller than the day-to-day variation in the Earth's atmospheric pressure that you hear about on the weather forecast. Another suggestion is that air tides in the high upper atmosphere may be significant in forming weather patterns.

More solidly established are **Earth tides**. The Moon raises a tide about 1m high at the Earth's Equator while the Sun produces one about half as big. This may not seem much in a planet which is 12,700km across. But this movement has been associated with the timing of volcanic outbreaks and earthquakes. US and Japanese scientists have studied the timing of both and found that when the solar and lunar earth tides coincide to form one major tide, a significant earthquake or volcano may be more likely.

But it is of course the tides in the Earth's oceans that are the most noticeable, and matter most, to humans and the rest of life on Earth. Let's take a closer look.

Tidal variation

Although no other body in the universe is close enough to raise a detectable tide on the Earth, even the tides produced by the Sun and Moon are complex. For one thing, you might expect the basic shape of a tide to be a bulge towards the Sun or Moon caused by their gravitational pull. In fact, the interaction between the Moon and Sun and the rotating Earth in effect stretches the Earth, plus the water and air around it, along the lines joining the Earth's centre to the Moon and Sun. This means that the Moon and Sun each set up two bulges in the oceans, the atmosphere and the Earth itself, one pointing directly towards them and another directly away from them. That is why you get two tides a day, not just one. The Sun tides are 12 hours apart, but the Moon's are about 12 hours and 25 minutes apart, the difference being caused by the Moon's movement around the Earth.

The height of the ocean tide on any particular day is determined mainly by the interaction between the lunar and solar tides – whether they are working together or in opposition.

The easiest way to think about it is to imagine that the two pairs of bulges raised in the Earth's oceans by the Moon and the Sun can either be aligned, or at right angles (at quadrature, in astro-speak). When they are aligned, the result is a very high tide followed by a very low one, as the

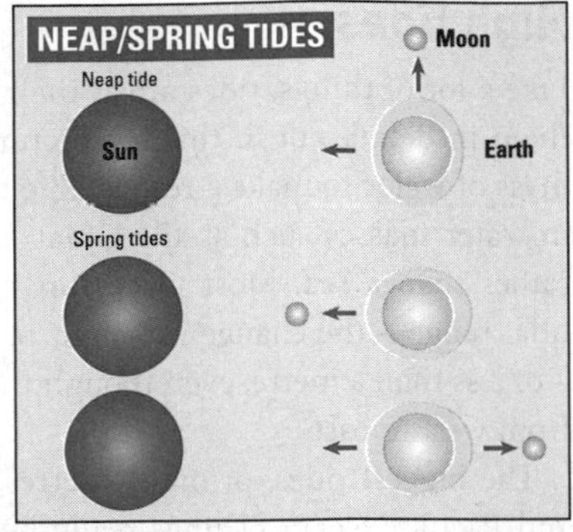

low-water parts of the ellipse match just like the high-water ones. This is a **spring tide**. It has nothing to do with the season associated with rabbits. Instead it gets its name from the way the water springs back and forth a long way when it occurs.

If the two bulges are at right angles, there are still tides because the solar tide is much weaker than the lunar one and is too small to counteract it entirely. But they are much lower, and are called **neap tides**.

As usual, there are further complications. As we saw above, the Earth's orbit around the Sun is not exactly circular and we are closest to the Sun in early January. So the solar tides at that time are higher than six months later, and can therefore go further towards cancelling out or reinforcing the lunar tides.

The Moon's orbit round the Earth is even more eccentric, 5.5 percent away from circularity compared to 1.7 percent for the Earth's orbit around the Sun. This means that its distance from the Earth varies by about 42,000km. The tides it raises at its closest (or perigee) are noticeably higher than those it produces when furthest away (at its apogee). Although gravity changes with the square of the distance between two objects, the tides one raises on another go with the cube of the distance, so this effect is bigger than you might expect.

In addition, although the Moon's orbit around the Earth is more or less aligned with the Earth's track around the Sun, the match is not perfect. In fact they are at an angle of about 5° to one another. That is why there is not an eclipse of the Sun at every new moon – instead, it only happens when the Moon crosses the line the Sun follows in the sky at exactly the moment when the Sun happens to be there. The result for our purposes is that sometimes the tidal bulges raised by the Sun and Moon are better-aligned than at others. Summary: during a January spring tide during an eclipse with the Moon at perigee, stand well back from the beach. But a neap tide in July, with the Moon at its furthest north or south of the Sun and at apogee, is going to be far less impressive.

High tides

Like a lot of things, tides are at their most interesting when you look at them in detail, not in this big-picture way. In practice, you need a big mass of water to make a respectable tide. Although they can be detected in water masses such as the Great Lakes, this really calls for an ocean rather than a sea. Most places on the Mediterranean coastline have a tidal range – the change in height of the sea between high and low tide – of less than a metre, even though the Mediterranean measures 3300km from west to east.

The biggest tides, of many metres, are raised in the Atlantic, Indian and Pacific Oceans. Things really become interesting when they arrive at the coastline. Because of inertia, the tidal bulges in the oceans never point directly at the place in the sky where the Moon or Sun can be seen. Instead they lag behind them. But when the sea reaches the land, the bulges lag even further behind. A large area of shallows near the land means that many cubic kilometres of seawater that were running along happily in the open ocean pile up and rub along the sea floor, losing energy and slowing up because of friction. And the lag gets even worse when a large landmass is in the way. Thus a high tide in the Scilly Isles, at the south-west corner of the British Isles, takes over five hours to get to the Orkney and Shetland Islands at the north-east. More intriguingly, it takes even longer to make the shorter journey to my home port of Liverpool. The reason is that to get there, the water has to make its way up the comparatively shallow Irish Sea.

Tides are complex but tide tables are accurate, partly because the position of the Sun and the Moon in the sky can be predicted accurately and partly because many years of data has been collected on the time lags in the system. A couple of centuries of tide data exist for major ports such as Brest in France.

However, another basic fact about tides is very simple. When lunar and solar gravitation push large volumes of water into small sea areas, the only way is up. The highest tides in the world are in the Bay of Fundy on the Atlantic coast of Canada and the Severn estuary in Britain (see box). Look at a map and you will see that the two places have something in common. Both have coastlines shaped to force the full tidal push of the Atlantic into a steadily narrowing funnel. In the Bay of Fundy, the tides are especially high because the water is built up in height by a strange resonance effect. In the Severn, the waters at their highest can form the **Severn Bore**, a wave – ridden by reckless surfers – that runs up the ever-

Top tides

Various points in the **Bay of Fundy**, Canada, claim to have the world's highest tides. This undignified squabble is mostly about dragging in tourists. My favourite river, the Mersey in England, has a tidal range of about 4m, which is pretty spectacular. Anywhere in Fundy has a range of over 12m, which would be a worthwhile sight, whatever the warring townships claim.

The Bay of Fundy in Canada may be misty, but it is home to the planet's biggest tides

There is little doubt that the biggest **total tidal range** – the variation in water level from the lowest low tide to the highest high – in the Bay of Fundy is about 15m. This beats a number of close contenders including Bristol on the River Severn in the UK (14.6m), Mont St Michel in France (12.3m), Puerto Gallegos in Argentina (13.2m) and Bhaunagan, India (12.2m). Also noteworthy is Derby, Western Australia, which is home to the biggest tides in Oz (11.8m), and whose town website is the source of these numbers (visit www.derbytourism.com.au and choose "Tides" from the "Derby" menu).

There is so much energy in tides that it seems a shame not to capture and use some of it. A **tidal power station** at La Rance in France has been doing so for thirty years, and there are others around the world. Some very large tidal power schemes, such as a plan for a massive barrier across the Severn between England and Wales, would have very severe environmental effects and have yet to happen, but there is fresh interest in the tides as a form of renewable solar and lunar energy.

narrowing river for miles, far from the sea, until it runs out of steam because of the friction of the river bed and banks. About eighty rivers around the world, from the Amazon to the Seine, have something like it (see box overleaf).

Over time, all this friction that occurs as the oceans are dragged against the shores by the tides means that energy is steadily draining from the

Biggest bores

It's a little like the breezy way people say "Hoover" when they mean "vacuum cleaner". Most people who think about tidal bores in rivers think about the **Severn Bore**. It is seen at its highest, with a good one being over 2m high, in Gloucestershire. But there are plenty of others. On the Trent in England there is one called the Aegir, named after a Norse water god.

The world's highest bore is called the Hangzhou and occurs on the **Qiantang River** in China, where surfers enjoy trying to stay on a wave that is anything from 7m to 9m high. The Mascaret on the **Seine** is much smaller, perhaps 2m. But it was as much as 6m high, second only to the Hangzhou, until river engineering works were carried out to modify it and reduce the damage it was doing. Also in France, the Garonne and Dordogne have notable bores. Perhaps the most persistent bore is that found on the **Pungue River** in Mozambique, which has been observed 80km inland.

Probably the most water is carried on the comparatively modest 2m high Pororoca – because it is a bore on the **Amazon**, the world's greatest river. Of other super-rivers, the Ganges also carries a bore, described by George Darwin, the British physicist and son of Charles.

For more information, see:

Tidal Bore Research Society tidal-bore.tripod.com/catalogue

Severn Bore www.severn-bore.co.uk

Earth–Moon system. One result is that the Earth's rotation is steadily slowing down. In fact days are getting longer by about 1.4 milliseconds (thousandths of a second) every century. That may not sound like much, but it adds up. About 600 million years ago, there were over 420 days in a year, with each day 21 present-day hours in length.

At the same time, the slow reduction in the energy available for the Earth–Moon system affects the Moon as well. It is edging gradually further away (at an average rate of 2.17cm per year) because less energy is needed to sustain a slower, more distant orbit. When there were 400 days in a year there were also 16 lunar months, in other words 16 lunar orbits round the Earth. At that time, the Moon would have looked larger and the tides would have been higher. Just now, the Moon has retreated to the point where it is the same size in the sky as the Sun – half a degree across – which makes it possible for a solar eclipse to occur when the Moon blocks it out. Come back in 620 million years, when the Moon has got even further away, and total eclipses will be a thing of the past. A few hundred million years ago, by the same token, they must have been more common.

Living with the tides

Life on Earth has had plenty of time to adapt to tides and they are integrated into the activity of many living systems. The tidal layers on the beach provide a series of different habitats to which plants and animals have adapted themselves. They are – from the top down – the splash fringe level, where water only splashes the shore rather than covering it; the high tide level, covered for several hours a day; the mid-tide level, covered about half the time; the low tide level, uncovered for a few hours per tide; the low fringe level, whose denizens can stand short periods out of the water; and, well, the sea.

Animals in the tidal zone know that the fall of the ocean means the chance to hunt for food if you are a predator. If you are prey, it is time to burrow or swim out of harm's way. Some forms of life have adapted

Varves

How do we know that there were over 420 days a year, 600 million years ago? By counting.

You only need walk in the woods from time to time to know that trees like the summer more than the winter. They grow faster when it is warmer and therefore their age can be determined by counting **growth rings**. But some creatures express other periodicities as well. Corals and other creatures that form solid shells grow in the hours of daylight, when they build up rocky material. At night they stop. Between one day's growth and the next, you get a slight band which can be observed with a microscope. There are also monthly growth bands – which can be harder to spot – caused by the rise and fall of the tides giving the coral more or less nutrition. And because the summer days are longer and hotter than days in winter, there are annual rings caused by the fact that the daily rings formed in summer are fatter than the ones laid down in winter. Look closely enough and you can count over 400 of the daily variety for each annual ring in a Cambrian shell from 600 million years ago.

Left to itself, nature provides all kinds of clues about its age. Take **varves**, for example. These lake deposits mainly encountered in Sweden tell us about the chronology of the past in a unique way. When the snow melts, spring floods wash large amounts of mud into the lakes, where it is laid down in dark layers of coarse sediment. Later in the year, fine, light sediment can settle because the water has become calmer. So a varve is a pair of light and dark sediment layers. Its thickness reflects the weather in a specific year so that – like tree rings – you can use them to build up lengthy climate records. Even better, the streams that fill up the lakes carry pollen and other plant debris with them so that the varve material also contains a record of the local ecology at the time.

specifically to conditions in the tide zone. The most startling example is the mangrove, a family of trees that can cope with salt water and to having their roots exposed at low tide.

Because of rising sea levels, scientists have been paying more attention to salt-resistant species. Sugar beet is derived from plants that grow on salt marshes and does not mind a salty environment. The same goes for barley. There is also a salt-resistant tomato that grows on beaches around the Pacific. The only problem is that it tastes terrible. And there are sea grasses, species of grass that can prosper in the ocean provided the water is shallow enough for the Sun to get to them.

Solar radiation

Take a look at the power output of the Sun, and it is obvious that **visible light** accounts for most of it. After all, the Sun looks yellow. We have evolved to make the most of the available light, which is why most animals have eyes that work in light wavelengths. Indeed the colour to which the average eye is most sensitive is known to optics as "visual" and is a sort of straw yellow close in hue to the colour of the Sun.

But the Sun bombards the Earth with far more than light. During World War II, a British Army group looking at interference with radar discovered that the interfering was mainly being done by the Sun, not by the Germans. This discovery was one of the cornerstones of the science of radio astronomy.

As our knowledge has grown, we have realized that the Sun emits a wide range of energy and particles. The first step on this road was taken in 1800 by the German/English astronomer **William Herschel**, better known, as we saw in Chapter 1, for discovering the planet Uranus. He used a prism to separate out sunlight into the spectrum, much as Isaac Newton had done over a century before. But then he placed a thermometer beyond the red end of the spectrum and noted that it rose. He had discovered infrared light – light below the red end of the spectrum.

We now know that the Sun emits **radiation** at all wavelengths we can observe, from the gamma, X and ultraviolet rays with shorter wavelengths than light, to light itself, and on into the longer, less energetic end of the spectrum with infrared and radio waves.

These invisible forms of radiation from the Sun are low in power compared to the light it emits. But they have plenty of important effects. The most striking is that ultraviolet light and X-rays from the Sun react with the upper atmosphere and produce charged particles called the **ionosphere**. This layer has the useful property of reflecting radio waves so that, with some cunning, it can be used for worldwide wireless transmission without satellites or subsea cables.

At the same time, the Sun generates a constant flow of particles, called the **solar wind**, whose composition more or less reflects that of the Sun itself. So it contains large numbers of protons and alpha particles (nuclei of hydrogen and helium) along with electrons.

All these particles carry electrical charges, and they emerge in especially massive numbers when the Sun is at its most active – when there are many sunspots, which are magnetic in origin, along with flares and mass ejections from the Sun. During extreme conditions on the Sun, some satellites in orbit around the Earth are shut down and turned away from the Sun to avoid damage to their electronics and their solar-energy-driven power systems.

These effects stay at a safe distance because the Earth's magnetic field (see p.121) diverts such charged particles away from the surface of the planet. This is just one more way in which the Earth looks after us. A planet whose surface was bombarded by the solar wind would be a far less inviting one.

Space weather

What happens when these particles arrive at the Earth is a stunning tale that goes by a number of names, of which the snappiest but perhaps least accurate is **space weather**.

The Earth's magnetic field is a powerful one and its effects are felt deep into space. When the incoming particles are about 100,000km from the Earth, they arrive at a feature called the **bow shock**, where the Earth's magnetism diverts them around the Earth. But although this is often presented as a one-way process in which the Earth's good magnetism fights off dastardly invading radiation, things are not quite so black and white. Were it not for the pressure of the incoming radiation, the Earth's magnetism would spread far deeper into space. The place where the incoming solar wind and the Earth's magnetic field are matched in strength is caused the **magnetopause**. Here, a hot layer of charged particles called the **magnetosheath** forms.

The upshot is that most of the solar wind particles stream about the Earth and head off into the outer solar system. The overall effect looks rather like air streaming round one of those odd helmets which cyclists wear to break time-trial records. The long tail at the back is called the **magnetotail** and

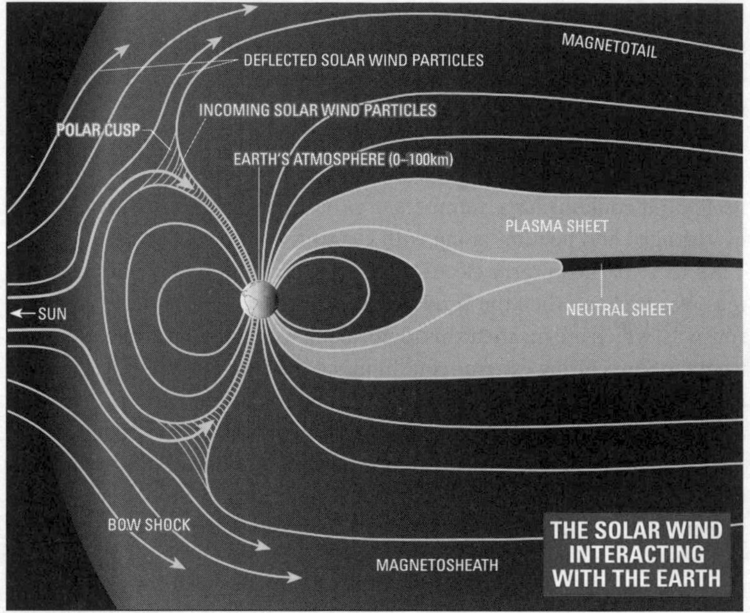

THE SOLAR WIND INTERACTING WITH THE EARTH

is many times larger than the Earth. It is detectable at distances far beyond any at which the Earth's gravitation could be discerned.

Some particles do, however, find their way towards the Earth's surface. They can do this because the **magnetosphere** (the bubble of magnetism that surrounds the Earth) is in fact far from spherical. Indeed, because the Earth's magnetic field is roughly aligned with the Earth's rotation (far more on this in Chapter 4), the lines of force that we use to visualize it can be thought of as springing out of the Earth's surface in the Arctic and Antarctic and looping round in space to join up. And charged particles like nothing better than a line of magnetic force to run along. When they do this, they end up close to the Earth's surface at regions called the **polar cusps**. Here, charged particles are fed into the **ionosphere**, a region of ions (charged atoms) and electrons in the upper atmosphere.

Many of these particles get spread out into two layers, the **Van Allen belts**, named after the US scientist James Van Allen (1914–2006). He discovered them in one of the first experiments carried out with satellites, in 1958. Because the Earth's magnetic field is not symmetrical, there is a spot in space above the South Atlantic (yes, they call it the **South Atlantic Anomaly**) where the inner Van Allen belt gets especially close to the Earth's surface and forms a hazard to satellites. The Van Allen belts are also a danger to people travelling in space, so space missions are designed to minimize the time that astronauts spend travelling through them.

On the ground, these effects are felt less often. But especially at high latitudes, tangible electric currents can be delivered to the Earth's surface, where they can pose hazards to electronic equipment. In 1989, the Canadian province of Quebec had its entire electricity system fail when such a current blew out the transformers. The more active the Sun is, the livelier all these effects become.

Lights in the sky

These charged particles from the Sun are also responsible for one of nature's most amazing sights, the aurora, which is also seen at its most dramatic during periods of maximum solar activity. It comes in two types, the **Aurora Borealis** in the northern hemisphere and the **Aurora Australis** in the southern, but that is a distinction without a difference as the two words apply to the same thing in different places. Both are manifestations of interplanetary particle physics visible to the naked eye (see colour section p.1).

The basic working of an aurora is simple enough. As we have seen, the charged particles which penetrate the Earth's magnetosphere generally end up in the Van Allen belts. But their confinement is finely balanced. If too many arrive at once, they overflow and hurtle inwards towards the north or south magnetic pole along the lines of force of the Earth's magnetic field. As they get lower, they meet the upper atmosphere and interact with the electrons in the air molecules far above our heads. The result is visible light. Aurorae (or auroras – take your pick) are common in the solar system. In 2005 they were detected on **Mars**, and they are also known to occur on **Neptune**, **Saturn**, **Jupiter** and **Ganymede**, one of Jupiter's satellites.

It is only in the past century that we have realized what an aurora is. It has long been known that major aurorae are associated with **magnetic disturbances**, shifts of the compass needle away from its normal orientation. So something electromagnetic is clearly going on. However, the idea that anything happening on the Sun could have such a direct effect on the Earth was regarded as science fiction until the twentieth century. In 1859 the British astronomer Sir Richard Carrington observed a bright flare on the Sun, but made no connection with the aurorae that followed, seen deep into equatorial latitudes, or the chaotic interruptions of telegraph traffic that ensued. Later on Lord Kelvin, a distinguished scientist whose word was more or less scientific law in Victorian Britain, published an article saying that any such connection was no more than coincidence. (Kelvin also "proved" that the Earth was 40 million years old, as we saw on p.17.) Key among the scientists who proved him wrong was Kristian Birkeland, a Norwegian whose struggles are described in Lucy Jago's *The Northern Lights*. Nowadays we accept easily that the number and intensity of aurorae we observe vary with the sunspot cycle.

The aurora is best observed along a belt (the annulus) of the Earth's surface that loops around the magnetic pole about 1500km from it. Any time you see an aurora, it is probably about 100km above your head. Its most characteristic colour is green, a wavelength of light created by charged oxygen atoms. Second favourite is red, also due to oxygen, and a blue colour from nitrogen is third. These colours are very precisely defined. If you look at the spectrum of auroral light, they appear as sharp lines. Each corresponds exactly to the amount of energy emitted when an electron around an oxygen or nitrogen atom that has been pushed into a higher-level orbit after absorbing energy from incoming solar particles

Geosight #1: The aurora

Although aurorae are visible far from the polar regions (real biggies have been seen in distinctly unpolar spots such as Hawaii), places such as **Scandinavia**, southern **New Zealand** and **Alaska** are favoured for viewing them. But don't go too far north or south. The region inside the annulus is rather poor as an aurora-watching zone. You'll see more of them a few thousand kilometres from the pole than you would at the pole itself.

Once you get there, the rules are simple enough: pick a time when the Sun is lively, and look upwards at night.

However, this is trickier than it sounds. For one thing, it is not for nothing that the polar regions are known as the **Land of the Midnight Sun**. In the summer, it is light most of the time. And the summer is the time when it is easiest to get to the far north or south, and when the temperatures are at their most acceptable.

Next, the aurora is a light in the sky. So you need a site where it has as little competition as possible. Somewhere dark and far from a major conurbation is ideal. You also need to choose a night as near as you can to the **new moon**, as a full moon will flood the sky with light. These are also the criteria for a good meteor-watching site (see p.48) so you may be able to spot both of these upper-atmosphere treats on the one visit.

Finally, you need the Sun to be active. The big flares that drive major aurorae are not predictable. But like sunspots, they go in an **11-year cycle** (to be exact, a two-peak 22-year cycle) that last topped out in 2000.

THE AURORA BOREALIS

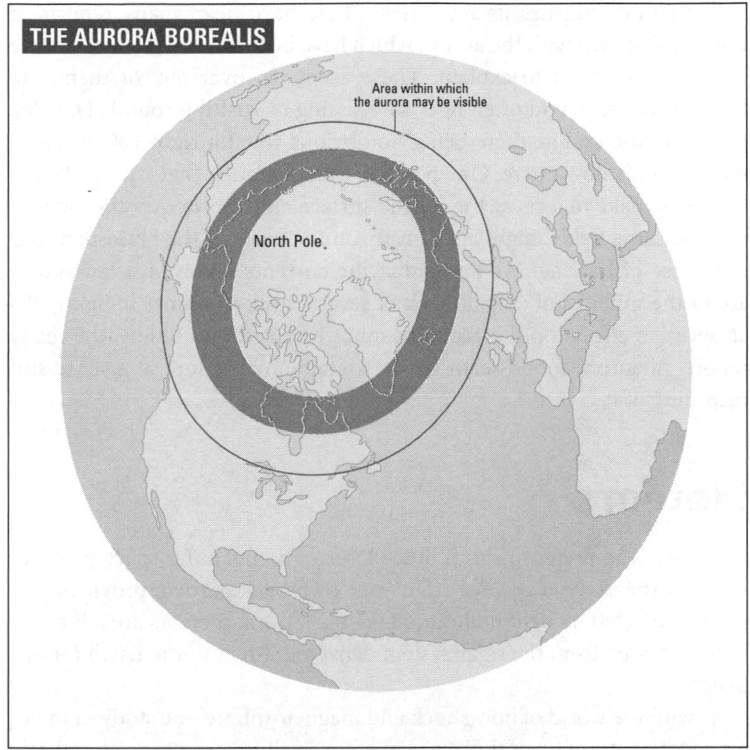

falls back again to where it started. The lower orbit needs less energy and the difference is emitted as light. As well as emitting light, the aurora can be detected with radar and can be used, like the ionosphere, for bouncing radio signals, a popular sport with the ham radio crowd.

Within the aurora there can often be seen shapes, described by the experts as arcs, bands, surfaces (which can be diffuse or pulsating) and coronas. These shapes can change rapidly (at up to 100km a second), with a rippling or pulsating effect, especially in response to major electrical storms. The resulting displays have been thought of as representing gods in battle and other mysterious likenesses. Their interpretation in Norse myth is especially gloomy, with suicide and violent death featuring prominently. Birkeland and others proved that inhabitants of the polar regions were wrong to think that the aurora could touch the ground. They are always far in the upper atmosphere. This is just as well, since being touched by the aurora was about the worst luck anyone could have.

The aurora still has its mysteries. There have been many reports of **noises** associated with the aurora which have been hard to dismiss despite being equally hard to explain. These accounts, over 300 of them, are highly consistent, and often refer to a hissing or rustling sound. The idea refuses to die despite there being no obvious way for light 100km up to make sounds down here. One possible explanation is that strong electrical fields make objects at the Earth's surface emit noise. Another idea is that the same fields might act directly on the parts of the brain that feed our sense of hearing. We know that the currents involved in an aurora are in the millions of amps (think of several million domestic lightbulbs at once) so effects on the ground cannot be ruled out. Oddly, the many reports of aurora sound seem not to contain any account of a successful recording of it.

Meteors

For those who prefer to stay a little closer to home, outer space provides lights in the sky you can see from your own back garden, provided it is not too affected by light pollution. Like the aurora, meteors are a little bit of high drama from the solar system delivered direct to the Earth for our delight.

As with the world of bow shock and magnetosphere, the study of meteors involves its own vocabulary. A **meteor** itself is something completely intangible. In fact, it is a streak of light in the sky, although they can also be detected by radar. Indeed, the invention of radar allowed meteors coming at us from the direction of the Sun, and hence hidden by daylight, to be observed for the first time.

A **meteoroid**, by contrast, is the solid object whose arrival in the Earth's atmosphere actually causes the flash of light. Most are about the size of the dust motes that you might spot in the air on a sunny day. But some are far bigger and if one is big enough to get to the Earth's surface in recognizable form, it is put in a collection and called a **meteorite**.

Meteors are an atmospheric phenomenon – the name has the same root as meteorology – that tell us about the universe around us. They come in two species. One lot are those that come in "**showers**". These are defined in time, maybe lasting a few weeks, or in some cases just a few hours. The rest are called **sporadic** and appear without any affiliation, at a rate of perhaps ten an hour on a clear, dark night. Some astronomers argue that there are no sporadic meteors, just shower members whose affiliation has not yet been identified or whose shower has become too diffuse to be observed distinctly.

What meteors demonstrate is that the solar system is not just an ordered array of planets running on predetermined tramlines round the Sun. Instead it is packed with overlapping flows of matter. Most meteor showers are caused by leftover debris scattered along the orbits of passing comets whose paths around the Sun intersect with that of the Earth. Sometimes the meteors are so faint that they are only observable with a telescope, but other showers are prominent features of the night sky. Perhaps the best-known is the weeks-long **Perseid** shower, seen from the northern hemisphere in August. Another favourite is the **Quadrantids**, a short, sharp shower, again seen in the northern hemisphere, that may be less familiar because fewer people want to sit up in the early hours of the morning in January than in August.

It is a tribute to the reach of modern science that the comets associated with all the major meteor showers have been identified. **Halley's comet** even has two, one in spring and one in autumn, since it crosses the Earth's orbit on its way towards the Sun and again on its way back.

Meteor showers

Here is a selection of meteor showers that are regarded as fairly reliable:

- **Quadrantids** 1–5 January
- **Lyrids** 16–25 April
- **Eta aquarids** 19 April–28 May
- **Southern delta aquarids** 12 July–19 August
- **Perseids** 17 July–24 August
- **Draconids** 6–10 October
- **Orionids** 2 October–7 November
- **Leonids** 14–21 November
- **Geminids** 7–17 December

Of these, the Quadrantids and Draconids arrive too far north in the sky to be seen readily from the southern hemisphere. The rest should be visible from anywhere in the world.

For more information, see:

Society for Popular Astronomy www.popastro.com

Any of these showers is worth a look, and seeing one is about the cheapest form of science you can do. The big thing you need and may not have is not a telescope or a satellite, but a dark sky. This means one free from street lights and the general murk of the big city, but also from bright moonlight. If there is a full moon in the sky, you may as well go back to bed.

If not, and the sky is dark, you can expect to see ten to twenty meteors an hour if there are no major showers going. If there are, you are in for a treat. But what is all this talk of Perseids and Quadrantids? Meteor showers are named after the part of the sky that the meteors seem to come from. Imagine (but be careful doing this for real) standing on a railway line. The lines seem to diverge out from a single point in the distance.

Is that a meteorite?

Few items in modest local museums are misattributed as often as alleged meteorites. If there were bogus Picassos about on the same scale, the curators would be fired en masse.

Meteorites and tektites sold by dealers are usually the real deal. (The same cannot be said of the wares peddled by some of the more imaginative fossil dealers.)

Meteorites come in two kinds. One are the **falls**, meteorites actually seen falling from the sky. The odds on your seeing such a sight are incalculably long, but it does happen a few times a year. If it does, the provenance of the rock in question – as the antique dealers might say – is pretty solid.

The other group are the **finds**. These are meteorites that are, well, found, either by chance or by searchers looking for them after reports of a fall. A fresh one might well have a fusion crust, a black outer layer from the heat of its passage through the atmosphere. An iron meteorite might well look shiny, and will be unexpectedly heavy for its size. But a stony one, although it will be darker than most Earth rocks, is more likely to be missed. This is why there are fewer stony meteorites among finds than among falls. The most distinctive are the chondrites because of the little spotty bits – the chondrules – that they contain.

Another clue is that meteorites tend to flock together in "**strewn fields**". So if you have found one, you might find another nearby.

Most of the bogus meteorites, incidentally, are in museums in areas where iron smelting has been a local industry. They are discarded bits of furnace slag.

Top tip – if you do find a freshly fallen meteorite, bag it up in something airtight. The gases it contains are full of scientific information but are lost after a spell in the open air.

Meteors are like that – members of a particular shower are moving in parallel and so seem to come from a point in the sky called the **radiant**. So the Perseids have a radiant in the constellation Perseus, and so on. (The constellation Quadrans Muralis was killed off years ago and the radiant of the Quadrantids is in Bootes.)

If there is a major shower running, just sit and look. A telescope will have too small a field of view, so go for what the military term the Mark 1 Eyeball instead. Don't stare at the radiant, but a few hand spans away from it. At the peak you may be seeing a meteor a minute and, depending on the shower, they could be slow and graceful or quick and easy to miss. It is worth staying up because as the night wears on, the Earth will be turning into the meteor stream and sweeping more up.

Even better is the treat of a meteor storm like those that come from the **Leonids**, a November shower, every 33 years. This is when the Earth crosses the part of the orbit that has been freshly seeded with new material, and there can be hundreds of thousands of meteors an hour. These big streams also prove that the universe we inhabit is no unchanging place. The big Leonid storms (they peak in years ending 33, 66 and 99) only began to be observed in the Middle Ages. The meteor streams in space that cause them have been mapped and it may well be that they will stop intercepting the Earth in a few hundred years, depriving our descendants of this friendliest of spectaculars.

Crater crazy

Of course, not all arrivals from space are so benign. In the past few decades, few changes in human awareness have been so striking as our acceptance that we might be abolished by **meteorite impact**. Part of the cause is the knowledge that the Moon is covered in impact craters. Another factor is the recognition of an increasing number of impact craters on the Earth, mainly in old, stable rocky areas like much of northern Canada. Now about 170 such structures are recognized on the Earth, including the first good candidate in the UK, spotted by oil prospectors in the North Sea below a kilometre of rock. The most visible is **Meteor Crater, Arizona**, a spectacular crater which also has a reassuring quantity of meteorites scattered around it. It is the biggest extraterrestrial site on the tourist map, not least because of its strategic location just off Route 66 in Arizona, although it fails to appear in the song by Bobby Troup. The biggest suggested impact is the arc on the east side of **Hudson's Bay** in Canada. Take a look in the

Impact craters on the Earth

It's a lively place, the surface of the Earth. The rain falls, the winds blow and the sands drift, and over a slightly longer period, the continents move about and recycle large chunks of the planet's crust. So **impact craters** may appear at a steady rate, just like they do on the other planets and satellites, but after a while, erosion means that they cease to be visible. In any case, most of the Earth is covered in water most of the time, so smaller impacts make a splash but do not leave a permanent record.

The best place to hunt for craters is in areas such as the **Canadian shield** where a flattish, rocky surface has been exposed to celestial bombardment for hundreds of millions of years.

In addition, all this hustle and bustle means that there are always several possible explanations for any crater-like object on the Earth's surface. It could be a **sink-hole** made by rain falling on limestone, or a leftover bit of a **volcano**, or a **bay** caused by the natural action of the sea. The one that I investigated during my university fieldwork, St Magnus Bay in Scotland, is now generally regarded as a sort of glacial/marine double act.

So any list of the Earth's meteorite craters should be filled with caveats. But the top ten more or less visible craters should probably look something like this:

1 Meteor Crater, Arizona the world's top tourist crater

2 Clearwater Lakes, Quebec, Canada a classic double impact feature like many seen on the Moon

3 Sudbury, Ontario, Canada a huge 60km by 30km structure about 1.8 billion years old and now mined for nickel and other metals

4 Lake Manicouagan, Quebec, Canada about 212 million years old and 70km across. Described by the local tourist board as "wild yet accessible"

5 Chicxulub, Mexico hard to see, like the dinosaurs it may have killed off

6 Ries Kessel, Bavaria, Germany "crater" is the Latin word for a cup but the Germans call it a kettle or cauldron

7 Henbury, Australia a set of craters about 42,000 years old

8 Roter Kamm, Namibia tough to get to but the satellite images are spectacular

9 Tenoumer, Mauritania same applies

10 Wabar, Saudi Arabia remote group of craters first described by British traveller St John Philby and now yielding interesting meteorite science

For a more comprehensive list of probable and possible impact sites around the world see www.somerikko.net/old/geo/imp/possible.

atlas – it is much the biggest exact arc of a circle anywhere on the map of the world and corresponds to a radius of about 400km.

There are various ways of finding out if a suspicious crater is the result of an impact. The most obvious is to look for bits of meteorite but this only works with the smaller specimens. Big ones are too badly vaporized to leave fragments behind.

Some possible impact craters tend to show up in areas of acute geological disturbance, which does little to enhance confidence in their authenticity. However, there are some geological structures that seem to be completely characteristic of meteorite impact, such as a distinctive rocky shape called a **shatter cone**, formed by the intense pressure of the impact, and a type of mineral called **shistovite**, which is a form of the common mineral quartz which has been altered by impact pressure. Some craters such as Mistastin Lake in Canada, and many craters on the Moon, have a **central peak**. Bigger ones such as the Clearwater Lakes, again in Canada, start to have a succession of **ring-shaped ridges** in which the rock strata have been lifted bodily by the impact like a car bonnet in a head-on collision. Other symptoms include a deep layer of shattered rock (called by the Italian word **breccia**) within the crater itself. Because this rock

Get your kicks just off Route 66: Meteor Crater, Arizona, is about 1500m across and about 50m deep. It has been there for about 50,000 years. The effects of local geology mean that it is noticeably square rather than round

is less dense than solid rock would be, gravity is detectably less strong above one – not so you'd notice, but a sensitive instrument can detect the anomaly.

Some meteorite impacts are even associated with "**tektites**", one of the oddest forms of rock found in nature. They are glassy masses formed by the impact and then flung far away, with such energy that they form a "**strewn field**" far from the impact itself. They tend to be called after the area in which they are strewn (moldavites, javanites, etc) and have distinctive aerodynamic shapes formed during their flight. The biggest weigh a few kilograms.

Mass extinction

Chicxulub, a massive crater in Mexico, is deservedly a household name despite its almost complete invisibility. Its discovery brought the ominous term "**mass extinction**" into the public imagination. It is almost certainly the crater whose creation killed off the **dinosaurs** and millions of other species about 65 million years ago. Exactly how the dinosaurs died out became the subject of speculation not long after they were first given that name, by British scientist **Sir Richard Owen**, in the 1840s. One theory was that they had died of constipation after changes in plant life. Intense radiation from a star near the Sun going supernova was another prospect.

Things changed in 1980 when a group led by the geologist Walter Alvarez started to find traces of the metal **iridium** at the top of the rock sequence called the Cretaceous (see p.95). Rare on Earth, iridium is often associated with meteorites. It turned up in various sites around the Earth at the same place in geological time, just the point where the dinosaurs vanish. Even then, it took time to establish that Chicxulub was the right age. This 150km crater, partly extending into the Atlantic off Mexico, had already been detected by its gravity anomaly.

This all sounds very satisfying, but just how does an impact in Central America kill off the dinosaurs in Europe? That too was an active research area in the 1980s because of fears of "nuclear winter", the prospect that the explosion of many nuclear weapons might alter the climate fatally. The basic mechanism is that the impact hurls vast amounts of dust into the atmosphere directly, and starts fires that add to the murk. The damage is done because sunlight cannot reach the surface, the temperature plummets below freezing and edible plants die off. This explains the end of the dinosaurs but also the extinction of other species too, which makes it better science than the constipation idea.

Does all this mean that the energy now being put into detecting and even diverting asteroids and comets that might strike the Earth is worthwhile? Well, maybe, and more on this in Chapter 9. The technology to do anything about an approaching NEO (Near Earth Object) is not in place yet. Maybe the hazard is tiny compared to humanity's other problems. But there is no doubt that it is real. In 1908, a comet head or perhaps a rocky meteorite struck a region called **Tunguska** in Siberia and devastated an area of many square kilometres. The blast was seen and heard hundreds of kilometres away. Such an impact would have killed millions if it had occurred over a major city, but it seems that on this occasion, only one person was killed, a reindeer herder thrown into a tree.

3

The solid Earth

The solid Earth

Now for the hard stuff, literally. The Earth is a solid ball, 40,000km in circumference, which translates to being 12,732km across. In other words, whether you dive deep in the ocean or climb Everest, the centre of the Earth will always be about 6366km below your feet.

A quick sum shows that this gives the Earth a surface area of 509 million square kilometres of water, land and ice. This chapter will focus on the layer immediately below the Earth's surface – the crust – and the forces that shape it. In the next chapter we shall come on to the deep Earth and its visible effects around us.

A shifting outer shell

Like any other crust, the Earth's is a stiff outer layer that conceals what lies beneath. However, it is not a monolithic structure that sits unchanging and impervious below our feet. It is being transformed the whole time and on every timescale, from the day-to-day change associated with minor earthquakes on up to the shifting of continents and the opening and closing of oceans over millions of years.

The idea of the Earth's crust has a solid foundation in evidence. Most of it comes from **earthquakes**. Although many earth scientists spend their careers trying to reduce the hazard that earthquakes pose to humanity, they also find them irresistibly interesting. Just as the true journalist loves a good murder for the front-page stories it will yield, so the Earth scientist appreciates a nice strong earthquake for the information it provides.

Even a modest earthquake registering 6 on the Richter scale releases as much energy inside the Earth as a one-megaton nuclear weapon. As the shock waves it produces spread through the Earth's interior and are detected across the Earth's surface, they provide information about the world beneath our feet. The approach is more or less a nature-powered version of the seismic technology used to hunt for structures which might contain oil and other resources inside the Earth. In addition, this talk of megatons is no coincidence. The same methods are used to detect underground nuclear explosions. Because armies tend to have bigger budgets than universities, one might argue that the Cold War was the best thing that ever happened to geophysics, the application of physics methods to the Earth.

These shock waves are able to tell us about the thickness of the Earth's crust. It is apparent from observing the time they take to reach detecting devices – called seismometers – around the world that some of the earthquake energy is being reflected back from a major feature several tens of kilometres below the surface. In addition, other energy is running along this subsurface join and then getting refracted back to the surface.

This marker at the base of the crust was first identified by Andrija Mohorovičić, a Croatian scientist, in 1909. It is known as the **Mohorovičić Discontinuity**, or the Moho. There are two main reasons why the join between two layers inside the Earth will reflect seismic waves strongly. One is a big change in the density of the rock, and the other is a change in the velocity at which sound will travel through that rock. Below the Moho, it seems, the rocks allow sound to travel at about 8 kilometres a second, a kilometre or two per second faster than at the base of the crust. This change is very abrupt. Not only does it cause some waves to be reflected back up to the surface, but it also means that those waves which do pass through the crust into the mantle will get to distant detectors faster than those that have come through the crust. This allows the thickness of the crust to be calculated.

Continents afloat

The surprises come when the distance to the Moho is measured across the Earth. It turns out that the idea that the continents float on the mantle is pretty accurate. Below the oceans, the Moho is near to the surface and the crust is at its thinnest. Here it can be as thin as 5km, with 12km more typical. Below a continental landmass, it is typically perhaps 45km deep with a maximum depth of 75km below major new mountain ranges such as the Himalayas.

This means that it is best to think of the oceanic crust as a raft afloat on a lake, whereas the continental crust is more like a big ship, or even an iceberg, where there is as much below the sea as there is above. The deep invisible parts of a mountain range are called its roots. As we shall see, there are many places on the Earth where the passage of time has left these roots visible.

Of course, one object will only float on another if it is less dense but more solid than the thing it is floating in. This holds for continents as much as for shipping. Thus the continental crust has a density of about 2.7 grams per cubic centimetre, which is the typical density of granite, its dominant type of rock. The ocean floor is dominated by the denser rock basalt, about 2.9 grams per cc, while the top of the mantle averages about 3.4 grams per cc.

It has long been understood that the continents and oceans do not start and end where the blue bits on the map meet the land. It was known that sea level was far from being fixed in any case, so the coastline was best regarded as a temporary demarcation. Instead, depth sounding had shown that at the edge of (say) western Europe, the sea gets steadily deeper until it reaches a depth of perhaps 100–200m. This area is called the **continental shelf** and is economically important for everything from fishing to oil production. Generally the shelf might be a few tens of kilometres wide but some extend for hundreds of kilometres.

But the edge of the shelf is rarely in doubt. It is marked by the **continental slope**, a steep plunge – at Alpine angles of up to 45 degrees – down to the **abyssal plain** thousands of metres below. This is the ocean proper.

During the 1960s, we began to appreciate the true difference between the make-up of the oceans and the continents, in a thought revolution as gripping as anything Einstein ever did for physics. This was the realization that oceans are new and constantly re-formed, while continents float about the Earth indefinitely.

Continents on the move

Even before the 1960s, there had been speculation for many years that the continents were not static. In particular, the apparent fit between the western coastline of Africa and the eastern shore of South America had tempted many observers to think they might once have sat together. Use the edge of the continental shelf as the marker instead of today's coastline and the fit is even more blatant.

The German geophysicist **Alfred Wegener** found many other lines of argument to corroborate the view that Africa and South America had once fitted together, especially from shared fossil species. He also looked back deeper into time and saw that there could once have been a single huge continent that he called **Pangea**.

However, Wegener died without seeing his theory gather a worthwhile evidence base. It was not until more than thirty years after his death on a scientific expedition to Greenland in 1930 that explanations began to emerge.

They started to appear with the development of instruments capable of measuring small variations in the magnetic properties of rocks. The centre of the action was the mid-Atlantic, which has long been known to be home to a ridge marked by active volcanoes and earthquake activity deep below the ocean. At some points, such as Iceland, it emerges above sea level. Measurements of the thickness of the crust along the **ridge** and other such ridges show that they are the places on Earth where the Moho is at its shallowest. It had been suggested that continents might drift because of the sea floor spreading from these ridges.

Now imagine sailing east or west from the ridge and measuring the magnetism of the rocks below. The scientists Fred Vine and Drummond Matthews found that the basalt of the ocean floor formed symmetrical bands of rock each side of the ridge whose magnetism alternated between pointing north and south. The reason (as we shall see in the next chapter) is that the Earth's magnetism periodically reverses itself, so that the iron atoms in cooling lava sometimes settle in one direction as it solidifies and sometimes in the opposite one. As everyone said at the time, the stripes were like the magnetic domains in a tape recording, although nowadays nobody still alive remembers what that was.

It is now possible to corroborate the steady opening of the Atlantic at a few centimetres a year (in a much-used analogy, about as fast as your fingernails grow) by reflecting light from lasers in Europe and America off satellites. The time taken for the light to arrive gives the distance with an accuracy of less than a centimetre and allows the widening of the Atlantic to be seen in real time. (See http://nercslr.nmt.ac.uk/nsgf.html for more information.)

But the steady generation of new material shoving Europe and Africa on one side of the ocean steadily away from North and South America on the other cannot be the whole story. For one thing, the Earth is not getting detectably bigger, so if there is a machine making new crust, there must be another one removing it. Areas where crust is being drawn deep into the Earth are called **subduction zones**. These are where old oceanic crust descends deep into the Earth as it is pulled below thicker layers of continental crust.

Geosight #2: The Icelandic Ridge

The Mid-Atlantic Ridge is normally hidden below about 2.5km of seawater, where it can only be explored with high-technology submersibles and other expensive gear (see www.lostcity.washington.edu). But for a more accessible view of ocean floor spreading in action, get on a plane to Iceland. Iceland is located on a hot spot in the Earth's crust (see p.111), meaning that far more lava is produced here than anywhere else on the Ridge, so much so that at this point it emerges above sea level.

On the island, there is constant volcanic activity and there are frequent earthquakes. The mechanism of ocean floor spreading is clearly visible, with volcanic eruptions causing long cracks to open up in the surface, which are filled by molten basalt. Along the plate boundary, which runs through the centre of the island from north to south, the rock is brand new, while at the edges, it dates back a few million years. But new vents emitting hot water and gases can erupt almost anywhere on the island, not just at the centre where the tourist spots such as Mývatn are. In the same way, the rest of the Mid-Atlantic Ridge is active across some undefined width, not just a narrow line of active volcanism.

The Mid-Atlantic Ridge at Myvatn in northern Iceland. To your left, America; to your right, Europe; and below, water being boiled by the heat beneath.

For the full picture we must enter the world of **plate tectonics** – tectonics being the term for the study of the Earth's structure on a large scale. Don't call this process continental drift in the presence of geologists, or you will sound like an amateur.

Sliding plates

The large areas of the Earth's crust that move about under the influence of plate tectonics are called **plates**. There are thirteen main ones on most reliable tectonic maps of the Earth. Tracing their boundaries is essentially about seismology (the study of earthquakes). At ridges where new material is being produced, small and relatively shallow earthquakes are common. They tend not to be massively damaging. At subduction zones,

Killer earthquakes

There is a grisly debate about just which earthquake has killed the most people, but there is little doubt that they are the most deadly of natural hazards, ahead of severe weather.

There are thousands of small earthquakes every day, and on average one big one a year at or above 8 on the Richter scale.

According to Stephen Nelson of Tulane University in the US, "earthquakes don't kill people, buildings do". He points to the biggest killer quake in recorded history, when some 830,000 people died in 1556 in **Shanxi**, China. They were living in caves cut into the soft blown soil called loess (see p.223) and were buried alive.

When **Tokyo** was devastated by an earthquake in 1923, much of the damage was done by fires which ran through wooden buildings. In the modern era, buildings in earthquake zones in the developed world are often very secure but the same cannot be said of the developing world.

However, the Boxing Day 2004 earthquake that affected **Indonesia** and other nations in south Asia and east Africa, with a death toll of almost 300,000, showed that it is harder to guard against the big waves (tsunamis) that a major subsea earthquake produces. A better early warning system that would allow enough time for evacuation is the preferred solution, but it will be hard to implement for the nearest communities because warnings would inevitably come only just before the tsunami itself. The near-total destruction of the city of Lisbon in 1755 was partly due to a tsunami which followed the initial quake, and Messina in Sicily suffered a similar fate in 1908.

Bill McGuire of University College London notes that the death tolls from natural hazards have been increasing, and attributes this to growing world poverty.

The table opposite lists the most deadly earthquakes on record, as compiled by the US Geological Survey. For more information on volcano hazards see:

Volcanism and Volcanic Hazards
www.ruf.rice.edu/~leeman/volcanic_hazards.html

earthquakes are often rarer and more destructive, as in Turkey or the very destructive Indonesian earthquake of Boxing Day 2004. At places where the plates slide past each other, energy again has to be released in earthquake form. Here the models are California, or the 1755 earthquake that caused the destruction of Lisbon.

A few major plates make up most of the Earth's surface, notably the Eurasian plate, which runs from the Mid-Atlantic Ridge to Japan, the

Date	*Location*	*Deaths*	*Magnitude*
23 January 1556	China, Shanxi	830,000	~8
26 December 2004	Sumatra	283,106	9.0
27 July 1976	China, Tangshan	255,000 (official) 655,000 (est.)	7.5
9 August 1138	Syria, Aleppo	230,000	unknown
22 May 1927	China, near Xining	200,000	7.9
22 December 856	Iran, Damghan	200,000	unknown
16 December 1920	China, Gansu	200,000	7.8
23 March 893	Iran, Ardabil	150,000	unknown
1 September 1923	Japan, Kanto (Kwanto)	143,000	7.9
5 October 1948	USSR (Turkmenistan, Ashgabat)	110,000	7.3
28 December 1908	Italy, Messina	70,000 to 100,000 (est.)	7.2
September 1290	China, Chihli	100,000	unknown
November 1667	Caucasia, Shemakha	80,000	unknown
18 November 1727	Iran, Tabriz	77,000	unknown
1 November 1755	Portugal, Lisbon	70,000	8.7
25 December 1932	China, Gansu	70,000	7.6
31 May 1970	Peru	66,000	7.9
1268	Asia Minor, Silicia	60,000	unknown
11 January 1693	Italy, Sicily	60,000	unknown
30 May 1935	Pakistan, Quetta	30,000 to 60,000	7.5
4 February 1783	Italy, Calabria	50,000	unknown
20 June 1990	Iran	50,000	7.7

African plate, which again starts at the Mid-Atlantic Ridge and covers all of Africa apart from its north-eastern extremities, and the Pacific, North American, South American, Indo-Australian and Antarctic plates, which you find more or less where the labels imply (see map opposite). Live in the centre of one of these plates and you are almost guaranteed not to die in a volcano or an earthquake. There are some seismically active faults, for example in Scotland, well away from plate boundaries, but they normally turn out to be fossil relics of old plate tectonic activity.

However, as the map shows, the picture is complicated by the existence of a number of much smaller plates, such as the Caribbean, Philippine and Arabian plates. One, the Juan da Fuca, deserves to be better known because its easterly edge, in California, is marked by the **San Andreas Fault**. This makes it perhaps the most closely observed geological feature on Earth. Its movement was responsible for the destruction of San Francisco in 1906 (see box), and it is constantly monitored for portents of what Californians term the Big One.

The existence of these small plates adds much to the supply of geological hazards. But it also tells us something of Earth history. As plates move around, new matter is either created at ridges or swallowed in subduction zones, characterized by deep ocean trenches. But it is impossible for this process to go on for hundreds of millions of years without some plates being broken up or partly consumed. Most of the small plates seem to have come into being in this way.

There are three main ways in which plates can meet: **creative**, **destructive** and **transformative**. Much as the Mid-Atlantic Ridge, pushing against the North and South American plates to the west and the African

The 1906 San Francisco earthquake

On 18 April 1906, at 5.12am, a major earthquake struck San Francisco. Up to 3500 people were killed and hundreds of buildings destroyed, many by the fires that broke out across the city after gas pipelines were ruptured.

The quake was caused by a modest-sounding movement of about 7m on the San Andreas Fault, probably in the Pacific to the west of the city. It was so large that local seismometers went off-scale, but it is now thought to have measured about 7.8 on the Richter scale. The shock waves from the quake did most of the damage by turning the soil underlying the city to liquid, a process called solifluction which occurs when soil containing water is shaken and turns to a liquid of porridge-like consistency. Today, the city's population is ten times what it was then, but safety standards for buildings have improved massively.

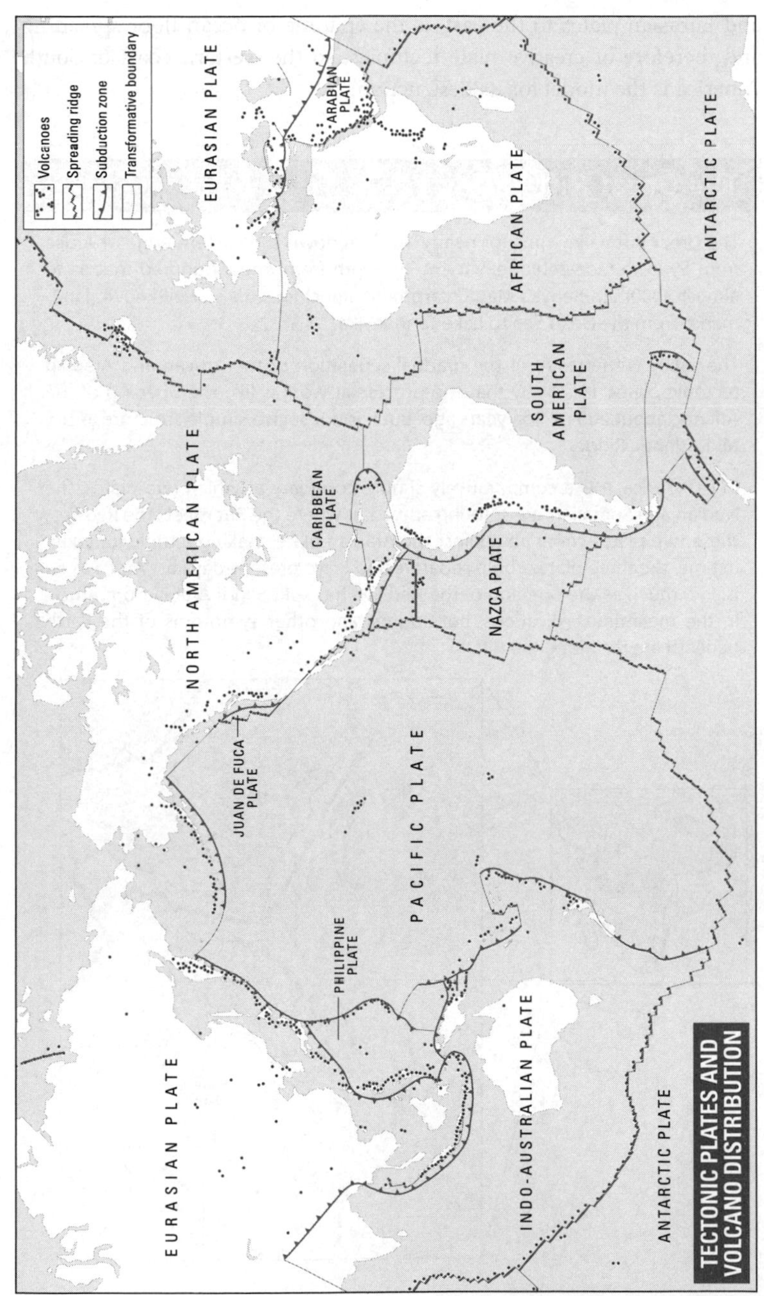
TECTONIC PLATES AND VOLCANO DISTRIBUTION
Volcanoes
Spreading ridge
Subduction zone
Transformative boundary
EURASIAN PLATE
ARABIAN PLATE
AFRICAN PLATE
ANTARCTIC PLATE
NORTH AMERICAN PLATE
CARIBBEAN PLATE
SOUTH AMERICAN PLATE
NAZCA PLATE
JUAN DE FUCA PLATE
PACIFIC PLATE
PHILIPPINE PLATE
INDO-AUSTRALIAN PLATE
ANTARCTIC PLATE
EURASIAN PLATE

and Eurasian plates to the east, is the epitome of ocean floor spreading and therefore of creative plate tectonics, so the western coast of South America is the model for its destruction.

The Great Rift Valley

The Great Rift Valley runs for nearly 5000km down the eastern edge of Africa from Syria to Mozambique. Varying in depth from a few hundred metres to almost 3000m in Kenya's Mau Escarpment, it includes many well-known landmarks, from the Dead Sea to Lake Tanganyika.

The valley is the result of the gradual separation of the African and Arabian tectonic plates. It is likely that the process at work is like the opening of the Atlantic about 200 million years ago. But it is not such a simple structure as the Mid-Atlantic Ridge.

In East Africa, it is a comparatively simple boundary at which two plates, the Nubian and Somalian, are splitting apart. But where the Rift meets the Red Sea, these two plates come up against the Arabian plate, making a triple junction. In time, the three plates could separate and leave present-day East Africa as an island, much as the opening of the Red Sea has split Saudi Arabia from Africa. In the meantime, volcanoes, hot springs and other symptoms of the Earth beneath are rife along the Rift.

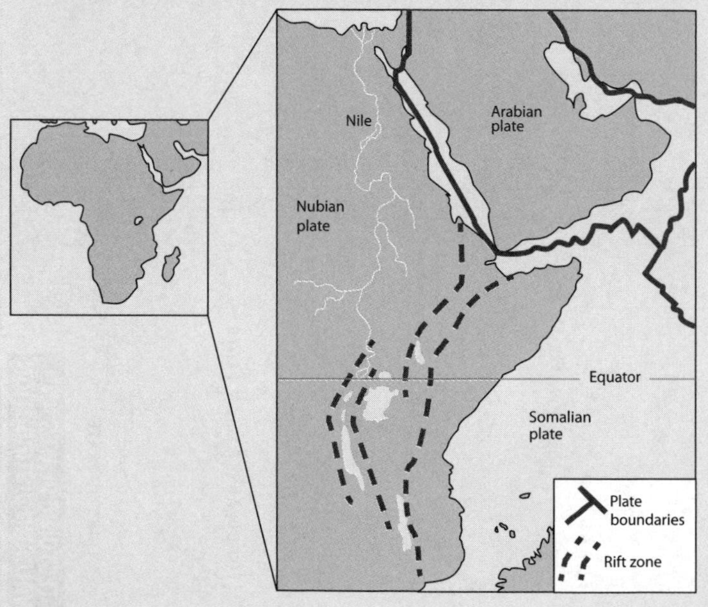

There, a deep ocean trench marking the edge of the Nazca plate is the spot where ocean crust is pushed below the South American plate, generating the earthquakes and live volcanoes of the Andes as it goes. The same process – this time where the Indo-Australian plate's northward movement has brought it up against the Eurasian plate – is responsible for the formation of the Himalayas, the greatest mountain range of today's Earth.

At a **transformative boundary**, such as the San Andreas Fault, the plates rub against each other in ways that can cause tremendous upheaval but involve only marginal gains or losses of crust area.

Although plate tectonic maps of the Earth look definite, not every boundary has been explored in detail. But the idea of the Earth's crust in motion turns out to have enormous explanatory power. The phenomenon of the "**Ring of Fire**", the band of earthquake and volcano zones that surrounds the Pacific Ocean, is accounted for. So is the fact that volcanoes associated with subduction zones produce lavas with a completely different composition from those responsible for sea-floor spreading. Indeed the Andes, produced as an effect of subduction, have given their name to Andesite, the lava that goes with it. Finally, the location of major mountain ranges is explained, as well as the forces behind their creation and destruction.

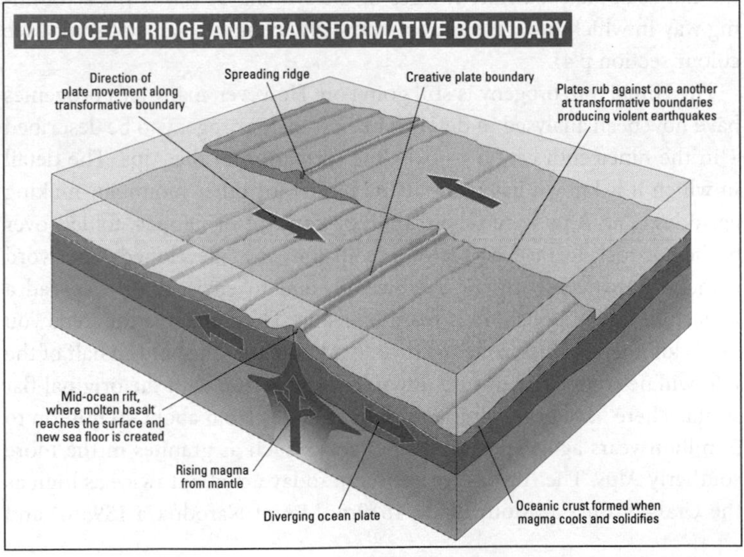

Making mountains

The mountain ranges we see today include some still under construction, others where the process has long ended and whose erosion reveals the innards of the mountain-making process, and yet others at every stage in between. The one thing we never see is a stable mountain range. Once it is built, a mountain is doomed to be removed, however long it takes.

Sometimes the force involved in raising mountain ranges is a strange one. For example, remember the idea of continents as ships floating in the mantle rock below. If the ship throws its cargo overboard, it will rise in the water. By the same logic, Scandinavia has been becoming more mountainous at a rate of a few centimetres a year since the end of the last ice age. It is springing upwards because of the weight of ice removed from it, a process called **isostatic rebound**. Nor is the effect confined to Scandinavia. Around Hudson's Bay in Canada, the isostatic rebound has totalled about 350m in the past 8000 years.

However, isostatic rebound is not a major means of mountain manufacture – or **orogeny** as the purists call it. The main method by which big mountain ranges such as the Alps, the Rockies or the Himalayas are formed is **collision**. The Himalayan orogeny – which has given us Everest (Chomolungma), the highest mountain of today's world at 8850m – is the result of the northward movement of the Indian plate as it crunches into the Eurasian plate. Satellite images of the Himalayas shows the astounding way in which they rise almost vertically from the plains of India (see colour section p.4).

The Himalayan orogeny is still going on. However, many past orogenies have now been analysed in detail. The first major orogeny to be described – in the nineteenth century – was the formation of the Alps. The detail in which it is known has made it the model for other mountain-making episodes. The Alps were where the key concept of **nappes**, folded-over bodies of rock formed by tectonic collisions, was developed. The word comes from the French for a tablecloth and to envisage one, spread a tablecloth out and push on it from one side. Then imagine the folds you see as kilometres-thick zones of rock. And note that the bottom half of the fold will be completely upside down by comparison with the original flat strata. There have been Alpine folding episodes from about 40 million to 5 million years ago, exposing deeper rocks such as granites in the more southerly Alps. The Alps as we see them today are about twice as high as the Urals in Russia (Mont Blanc 4810m, Mount Narodnaya 1895m) and are less than 200km from south to north, the direction of most of the so-called overthrusting that forms the nappes. But measuring the nappes shows that the folding has used up about 200km of crust. Look at the Alps today and you also see vast areas of newer sediment caused by the erosion of these big new mountains. Come back in a few hundred million years and you may well find something like the Urals, or even the mountains of Scotland, as we see them today, with these rocks gone and new ones exposed that today are many kilometres below the surface.

The world's biggest mountains have been formed either by such collision processes or at subduction zones, where material is vanishing into the mantle. This process produces deep ocean trenches such as the Marianas in the Pacific or the trench running along the western side of South America. But the area behind the trench also becomes compressed as all that ocean floor slides beneath it. The **Andes** are the result of this compression, enhanced by the new molten rock formed from crust material dragged deep into the Earth. Hence all the volcanoes and earthquakes in that region.

A good example of this process is the formation of the **Urals**. Politically the Urals are regarded as the boundary between Europe and Asia. But there is geological reality underlying this convenient line on the map. They were formed as a long chain of mountains over 2500km long but only 150km wide at most during the late Palaeozoic era, when plate tectonic movement closed the ocean between Asia and the predecessors of Europe.

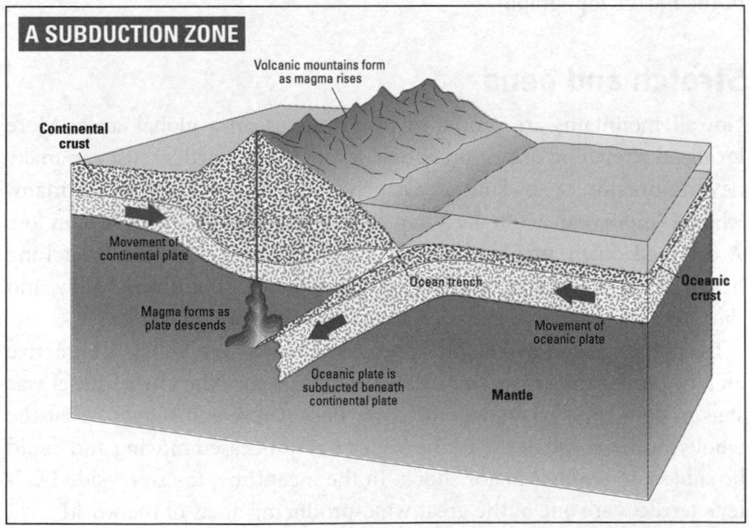

The detailed mapping of the processes responsible for the formation of the Urals shows that it began with a third continent roughly equivalent to present-day Kazakhstan. The western edge of this continent was active – picture volcanic islands offshore – while the edge of Europe was passive, and the ocean between them closed as the ocean floor was sucked below Kazakhstan and crust material piled up to mountain height. Only later did the Asian plate arrive to complete the story.

All this happened so long ago that the Urals we see now are essentially the roots of the huge range that was first thrown up. This means that we can see igneous rocks such as **granite** which were produced by melting and whose large crystals show that they solidified deep in the Earth. The chemical composition of granite is the same as the lava andesite but, having cooled at or very near the surface, andesite has far smaller crystals. The presence of metamorphic rocks – sedimentary rocks altered by the heat and pressure of the deep Earth – are another sign that the Urals as we see them today are the remains of a range of mountains that must have been of Alpine proportions when new. These revealed depths also house many valuable mineral veins, including ores of metals such as gold, silver and platinum.

Don't forget that these orogenies are not just local events. The **Downs**, a chain of beautiful hills across south-east England, were created by the Alpine orogeny, but at a distance where its power was far less than at its centre. The same applies to many of the gentler but still interesting parts of the Earth's topography.

Stretch and bend

Not all mountains are produced by processes on a global scale. More localized stretching or compressional forces in the Earth's crust can make new mountains, or lowland areas such as the Rhine Valley in Germany, which is more than 200m deep in places. The names here are German too. A dropped-down area bounded by faults – essentially where stretching has made the crust give way – is a **graben**, as with the Rhine Valley, and the opposite, where compression has raised an area, is a **horst**.

Despite its placid and stable appearance, the Rhine Valley is an active area of the Earth's crust. Towards its southern end, the city of Basel was shaken by a large earthquake in 1356. Recent research suggests that the whole southern end of the Rhine Valley has not ceased moving and could be subject to another major shock. In the meantime, the steep-sided valley's terraces are one of the great wine-producing areas of the world.

In addition, isostatic effects like those that caused land to rebound after the last ice age can kick both ways. Thus a large area of the Earth's crust that is subject to significant amounts of deposition of sediment can subside as the weight builds up. There can be faulting, where all or part of the structure slips en masse, but it is also possible for the whole lot to bend so that folding of the rocks is seen rather than faulting.

Another handy concept is **epeirogeny**, the opposite of orogeny. It involves the shifting of large amounts of crustal matter up or down without significant horizontal movement like the folding that makes major mountain chains. The resulting terrain is largely free of faulting except at the edges. This large stable area is known as a craton. Epeirogenic episodes seem to be associated with changes in the underlying mantle. The term is only applied to large areas of land or of the ocean floor.

The rolling hills of South Africa are a classic epeirogenic landscape. Little changed in millions of years, they have been created by successive episodes of uplift separated by long periods of erosion. The result is that the surface has been gradually scoured away by wind and water, and about half a dozen successive land surfaces of differing ages, known as erosion surfaces, can be identified. The process has had big economic effects. Millions of years of erosion have eaten away so much material that diamonds, formed deep inside the Earth, can now be mined there (see p.109).

Washed away

The one thing we have already found out about mountains is that at the Earth's surface, nothing is for ever. Eventually they are all worn down to the sea, a truth that poets have often found useful when on the hunt for a gloomy metaphor.

This process is **erosion**. This scientific term has had the rare luck to enter popular usage with its technical meaning more or less intact, unlike, say, a quantum leap, or schizophrenia.

Some erosion is simply caused by gravitation. Material is forever looking for ways to get downhill. Soft material in particular tends to "**creep**" downslope and will eventually find its way to a river or stream as it does so. In soil, creep can produce hillside terraces that are easy to mistake for the relics of human agriculture. But this unaided creep is a slow process. The main factor responsible for more appreciable erosion at the Earth's surface is **water**, and it comes in several forms including ice, rain, rivers and the sea.

Each of these forms of water is a hazard to the solid Earth, but in areas where it freezes in winter, **ice** is the biggest threat. When water runs into a tiny crack in a rock and freezes, it also expands. The effect is just like putting a bottle of wine in the freezer to cool it fast and then forgetting about it – the solid glass or rock cannot resist the expanding ice and ends up cracking. This can speed up the comparatively gradual process of water-based erosion massively.

Rain falling straight from clouds to the land is perhaps the least significant cause of erosion. However, it may be the principal erosional factor at work on the steep slopes of a high mountain where there are no rivers or glaciers at work.

When rain gets loose on a soft surface, it removes matter and shifts it downhill, a process called **rainwash**. Once the rain has formed into surface water, the forces involved increase and the phenomenon becomes known as **soilwash**.

At some point, rain that falls on land will find its way to a **river**. Rivers are often described as "cutting" the valleys through which they flow, but this is misleading. The river directly cuts the central channel in which it flows, but the rest of the valley is cut by smaller tributary streams or by water running across the land to reach them.

Even the way a river erodes a landscape is not a straightforward process. Thus, the Niagara Falls (see box) have been eroding their way backwards up the Niagara River since the end of the last ice age. In 12,300 years they have covered 11.4km, or 93cm a year.

Charles Lyell, the British geologist who determined the rate of recession of the Niagara Falls in 1841, would not agree completely with the current model of how rivers perform erosion. He was the father of **uniformitarianism**, the idea that the processes we see in action on Earth now are the ones that have shaped it through time. He was completely right about this and uniformitarianism is one of the great concepts of science. But we have a growing awareness that most of the erosion that a river carries out is not a slow, creeping effect but occurs in a few short episodes when the river is in spate. This is when it has the most erosion power, partly because the water itself is more abundant and is moving faster. In addition, especially in mountain streams, water that moves faster sets rocks and pebbles in the river moving and they erode the riverbed even faster.

Chemical weathering can speed the process of erosion. Its most notorious form is **acid rain**. Loaded with oxides of sulphur and nitrogen, this polluted water dissolves rocks such as limestone, as well as killing trees and harming fish and other wildlife. But the most common form of chemical weathering is **carbonation**. Here, carbon dioxide combines with rainwater to form an acidic solution that reacts with calcium car-

Geosight #3: The Niagara Falls

The Niagara Falls are one of the most viewable and well-visited geosights in existence. They stretch across the Niagara River between the US and Canada, interrupted by Goat Island. Just over a kilometre wide, they carry 100,000 litres of water a second. They are over 50m high, and are the biggest barrier on the short river which flows from Lake Erie to Lake Ontario. They have been there since the end of the last ice age and have existed in their present form for about 5500 years since water scoured out an old riverbed and created the distinctive whirlpool seen today.

As we saw on p.74, the falls have been eroding their way upstream since the end of the last ice age. In the last few hundred years this rate of recession has increased to around 1–1.5m a year, and, left to itself, the process would accelerate further in the future when the falls wear their way back to some softer rock upstream. However, engineers are working to slow this erosion, and have managed to reduce the falls' retreat to only about 30cm a year. Without this intervention, the falls would eventually work their way back to Lake Erie and cease to exist in about 50,000 years.

There is a wealth of Niagara information on the web. One of the better sites is:

Niagara Falls Live www.niagarafallslive.com/Facts_about_Niagara_Falls.htm

The Canadian Falls: just one part of the Niagara Falls complex

bonate in rock such as limestone to produce calcium bicarbonate, which is soluble in water. The result, in time, is **caves** (see p.196) – wherever there is limestone, there are caves, and people with wetsuits and head lanterns to make the most of them. (See www.showcaves.com to find some near you.)

The presence of limestone caves also reminds us that not all rivers are on the surface. Indeed **underground rivers** carry large volumes of water

UNESCO's geoparks

For many years, the UN Educational, Scientific and Cultural Organization (UNESCO) has been putting important places around the Earth on the **World Heritage List**. In 2006 there were 830 of them. Most are on the list for their role in human development: nothing wrong with that. But some are there for their natural attributes, such as the **Great Barrier Reef** in Australia, and the **Galapagos Islands**, politically part of Ecuador. In addition, another list of protected sites concentrates on **Biosphere Reserves**, mainly places with spectacular wildlife.

Now UNESCO is starting to name "**Geoparks**", sites of special geo-interest, around the world. To help them on their way, here are a few places that might feature.

In the Antarctic:

- **Mount Erebus** (see p.207) a live volcano now endangered by being on the tourist map
- **Dome C** the world's darkest and clearest skies
- **The dry valleys** (see p.207) just what the name implies, the nearest you can get to Mars without a rocket
- **The under-ice lakes** (see p.210) the world's most unexpected environment

In Latin America:

- **Lake Titicaca** the world's highest commercially navigable lake
- **South Georgia Island** because of the direction of the ocean currents, a detached bit of the Antarctic near to South America
- **Galapagos Islands** a Hawaii in miniature, featuring live and extinct volcanoes

In North America:

- **Major volcanoes of Washington State and region** Mount St Helens (see p.117) and its more tranquil allies
- **Most of Alaska** needs protecting from oil drilling
- **Deserts of the south-west** unique archaeology as well as ecology
- **Chesapeake Bay** the largest estuary in the US
- **Standbys such as Old Faithful, Niagara Falls (see p.75) and the Grand Canyon**

and perform significant erosion. My favourite place to see them emerging into daylight is the Jurassic coast of southern England, whose soft cliffs are in almost constant motion because of streams making their way through the mud and clay. The water eases the material onto the beach only for the next high tide to make short work of it. This is a sight not to be missed, if only because of the fossils it brings to light.

In Europe:

▶ **Jurassic coast of England** (see p.95) spectacular scenery, and the birthplace of scientific palaeontology

▶ **Burnt-out volcanoes of the Auvergne, France** now safe to visit but still a top place to see what volcanoes look like

▶ **Iceland at large** (see p.63) a bit of mid-ocean ridge we can all visit

▶ **Glaciers of Norway** may well be there when the Alps have lost theirs

▶ **Volcanoes of Italy** Etna, Stromboli, Vesuvius and many others including Vulcano, which gave its name to the whole phenomenon

In Africa:

▶ **Natural nuclear reactors of Oklo, Gabon** (now long inactive; see www.oklo.curtin.edu.au)

▶ **The Great Rift Valley** (see p.68) the future Atlantic

▶ **Glaciers of Kilimanjaro** (see p.233) preserving them will mean we have got a grip on climate change

In Asia:

▶ **Mount Fuji, Japan** (see p.114) under severe visitor pressure

▶ **Arctic Russia** needs proper environmental protection

▶ **Everest** the main routes are a rubbish tip

Elsewhere:

▶ **Reefs and volcanoes of Hawaii** (see p.111) unique mid-ocean volcanology

Many of these already have some sort of protection but all the evidence from other UNESCO citations is that its approval warns national governments, and other developers, of the potential for international uproar if they get it wrong. Let UNESCO have your ideas for the list.

UNESCO World Geoparks www.worldgeopark.org

Wave power

The sea and the erosion it carries out are perhaps the most striking reminder for many people that Earth forces are not something gradual and far-off. Most people never see a volcano, but if you holiday at the seaside at the same place each year, you are almost bound to notice change.

Many coastal areas of the Earth are plainly deposition or erosion coasts. Thus the British Isles are gradually tilting, with the west rising while the east falls. In the west, rivers need to be dredged and new land appears. But to the east, it vanishes, and as with rivers, most of the erosion happens during a few violent storms, not at a smooth, gradual rate.

Often the effects of marine erosion are spectacularly obvious. The erosion itself can occur as a slow sliding process or via unmissable avalanches or rock falls, with the more high-profile processes concentrated in areas of harder rock. Look at a cliff and you rarely see a smooth rock face. There are commonly steps in and out where layers of harder or softer rock have eroded at different speeds, or marking old sea levels where erosion was once severe but which have now been left far above high water. High cliffs, offshore stacks that were once part of the land, and arches connecting isolated rocks to the land are all signs of erosion. Come back in a few hundred years and some will have collapsed.

On softer and less rocky terrain, such as Chesapeake Bay near Washington DC, the signs are different but the result is the same. Here land is being submerged, partly because of tectonically driven sinking, so islands vanish and coastlines shift inland.

Wind erosion

The other Earth process that shifts material is the **wind**, which is a major contributor to erosion in dry areas such as deserts and the Arctic and Antarctic. Where wind speeds in a desert rise above a certain threshold level, the wind can pick up small, loose particles and carry them along. These particles in turn act as an abrasive, wearing away at rock surfaces. Over time wind erosion produces astonishing desert landscapes, consisting of large, scoured areas. While sand and sand dunes are the characteristic landform of deserts, most of the land area of most deserts is not sandy. Instead it is pebbly, because the wind has removed all the smaller particles of material from the rocks.

However, water, in brief **flash floods**, is also a major factor in forming dry landscapes. Although such floods occur only once every several years, they alter the landscape very severely when they do occur because the water runs over large areas of loose, erosion-prone material.

Deposition

All this erosion means radically altered landscapes. But because the Earth is a closed system, it also means that material is being dumped elsewhere. Once the ice, wind or water carrying eroded material away from its original location no longer has the energy to transport it further, the process known to geologists as **deposition** occurs.

If a glacier dumps material, it creates chaotic-looking terrain which we shall examine in more detail in Chapter 7 (see pp.221–23). More organized and satisfying to the eye are the landforms produced by the wind (**aeolian**, in the jargon). Where wind speeds fall below the level at which the wind can carry the material that has been eroded away, it is deposited, usually in **dunes**. Dunes typically form in long lines at right angles to the wind. In coastal areas, they are sometimes crescent-shaped because of the influence of onshore winds. The wind can push dunes along at a rate of several metres a year, as they are made of unconsolidated sand.

As most erosion occurs under the influence of water, most rocks are created by the deposition of material from water. As with the wind, the sediment-carrying power of water goes up rapidly with its speed, and falls as fast if it slows. Look at any river taking a bend and the effect will be apparent. On the outside of the bend, you will see the water moving at its fastest, and eroding the river bank as it goes. On the inside, where the water is slower-moving, you will see material being deposited. The deposits could be anything from chunky boulders to fine sand, depending what the river has flowed through on its journey so far. Such banks are also fine spots to find fossils and washed-down archaeological remains. Over time, the combination of erosion at the outer bank and deposition at the inner one will shift the watercourse across the valley. Many a tonne of concrete has been poured by civil engineers in attempts to keep rivers from shifting in this way.

The general rule that slower water means deposition also applies when rivers meet, with the smaller river often slowing and forming **sandbanks** and other structures. This process is seen at its most spectacular when a river meets the sea. The load of sediment it drops can form sandbars, shingle bars or small offshore islands that form a significant hazard to shipping. As the area just offshore is a high-energy environment of waves and surf, these structures are often eroded about as fast as they are built up, so an average sand grain in one of them may not stick around long before moving on.

When larger rivers arrive at the sea, they can form a **delta** (see p.188). The first delta to be described was that of the Nile, which carries exceptionally large amounts of sediment to the Mediterranean. Over time the river has deposited material hundreds of metres deep and steadily cut new channels called distributaries to keep making its way to the sea.

However, even a river delta may not suffice to absorb the sediment borne by a major river. Often it forms barrier islands, long islands parallel to the coast, from material which has been carried down the river and then been slowed to a halt when it reached the sea. As the name suggests, these are valuable for protecting the coast and the people who live there from severe storms. When conditions get rough, ships take refuge on their landward side.

Deposition in the sea is rarely a simple process. In high-energy areas with large waves and storms, rocky material such as pebbles tends to dominate. But if there is no such material available from nearby erosion, the coast will tend to be eroded back. In quieter waters you can get sandy or even finer silty deposits, perhaps on a lee shore with no big storms, or when there are offshore islands, reefs or sandbanks in place to keep the water calm.

However, deposition on an energetic shore is not for ever. Instead, it is subject to "**longshore drift**", the process whereby material keeps moving along the coast in the general direction of the wind and current. All those groynes put up along bathing beaches to keep the sand in place are there to slow this process down. That is why the sand builds up on one side of them, in the direction of its travel, and you get a large drop on the other side.

The deep oceans are also the site of deposition, especially the **abyssal plains** that dominate the deep ocean beyond the continental shelves, typically 4–6km below sea level. In areas such as the North Atlantic, where the ocean is opening and ocean floor is not being destroyed, the abyssal plain can build up kilometres of fine sediment. Some of it is the remains of life from nearer the surface, as bits of everything from plankton to whales rain slowly down. More significant is the sediment that arrives at the plain in massive subsea flows running down the continental slope and across the plain. The resulting deposits are called turbidites, aptly because of the turbid nature of the currents that emplaced them.

New rocks from old

If this deposited material remains undisturbed, over time it may eventually solidify to form **sedimentary rock**. Similarly, some of the material that is dragged deep below the Earth's surface by tectonic processes will reappear at the surface as **igneous rock**, recreated entirely by melting, or as **metamorphic rock**, severely altered by the heat and pressure of deep burial but not remelted and so retaining some of its original nature.

Let's look at each of these processes in turn, and examine the different rocks which result from them.

Sedimentary rock

For sedimentary deposits to be formed into solid rock, a cementing process called **lithification** must take place. In sandstone, the cement is typically iron oxide. Indeed, the silicon dioxide that makes up the grains is white – it is the iron oxide that makes the rock look yellow, brown or red, depending on how much of it is present. The iron oxide is deposited from water percolating through the sand once it is buried in the Earth.

But there are other processes at work in the process of lithification. One is the sheer pressure to which a sediment is subject as it is buried. This big squeeze is especially severe with muddy sediment. As it is compressed, the grains line up at right angles to the pressure. This is why shales and mudstones, rocks formed from mud, have visible layers which the mud does not have.

There are many kinds of sedimentary rock. Thinking of them in order of grain size is an informative approach. The finest-grained are the **silts** and **clays**. Here some geologist jargon comes into play. They will talk about sands, clays or silts when they mean hard lumps of rock made from these soft materials. In the same way among the volcanic rocks, they will describe a rock as a lava or an ash long after it has cooled and solidified.

The fact that such small-grain-size sediments can be deposited – and not washed away later – is virtually proof that the sediment formed in a low-energy environment. The larger-grained **sandstones**, by contrast, are made of sterner stuff, and come from a more energetic background. Sand itself consists mainly of silicon dioxide, in its most common form, **quartz**. Quartz's simple composition and its presence in most rocks make it pretty much the ur-mineral in petrology, the science of rocks.

The sand dunes you see in pictures of the Sahara or other major deserts today are replicated in rocks from periods such as the Triassic, exposed across large areas of Britain. Careful fieldwork will reveal the direction and strength of the wind at the time they were deposited, and the type of rock that was eroded to form the sand.

Sandstones are common in the Earth's crust because almost any rock, when it is eroded away, has the potential to produce sand. **Shales** are even commoner but because they are softer, they rarely form impressive cliffs or other exposures. Sandstones and shales are called clastic rocks, meaning that they are made up of clasts – the posh word for broken-down pieces of other rock – which have shaken loose as weaker parts of the rock were eroded away and have now been reassembled into new rock.

In honour of their colours, shales and sandstones usually appear on geological maps in shades of grey and yellow respectively. But on any map with a big swathe of sedimentary rocks, you will also see a lot of blue. This is the colour for **limestone**, which in the right light can indeed look blue and which is the main non-clastic sedimentary rock.

Limestone consists almost entirely of calcium carbonate in the form of calcite. It differs fundamentally from sandstone and shale because it is formed from living things. Limestones contain abundant fossils and it is possible to argue that they are almost fossils themselves. The most frequent fossils for them to contain are the hard parts of corals, which suggests that most have formed in warm, shallow water like those desirable holiday destinations which have coral reefs today. Other, much finer, limestones form from marine algae. Yet others are made from the shells (ooliths) of microscopic marine animals and are called oolites – pronounced oh-oh-light. Others consist mainly of larger pieces of shell. Very fine-grained limestones are called **chalk**, and ones with a substantial percentage of magnesium in their chemistry are called **dolomite**.

The grain sizes of limestones reflect their history. If coral has been allowed to die and settle to the seabed to be lithified, the result can be a rock with massive coral lumps, terrific for the head office of a bank. If the bits have been ground fine to ooze, a far-smaller-grain limestone is the result and a microscope is needed to determine the animals that make it up. (Yes, I appreciate that coral is a symbiont, not an animal, but you get the idea.)

These categories look clear enough on paper, but the Earth does not like things simple. Thus a limestone might be sandy, or a sandstone may be muddy, to the geologist's eye. In addition, components that make up only a fraction of a rock often have significant structural importance. While

Life makes rocks

If the Earth were completely lifeless, the rocks that make it up would be quite different from those we know. Although the Earth has produced life, life has also produced the Earth we see around us. The importance of life in making rocks is most apparent in the case of **limestone**, the various types of which are all derived from one or other form of marine life. But limestone is not unique in having its origin in living matter.

In the geological record, the first definite signs of life are found in the Precambrian. While the creatures of that era were simple and less diverse than later species, they did build impressive structures. The best-known are **stromatolites**, which in Australia and elsewhere have left fossil beds many metres thick. They are generally accepted to be the remains of thick mats of material produced by algae in shallow seawater. Their descendants are still with us but have declined in importance with the arrival of true plants and animals.

At the same time, virtually all the economically important minerals we use are derived from living creatures. The most obvious are the **hydrocarbons** – oil, coal and gas – which are derived from rotted plants. The whole global warming issue really arises because they were laid down over many millions of years and we are releasing the carbon they contain in just a few hundred, while being amazed that this has any effect on the Earth.

Despite the introduction of fertilizers derived from atmospheric nitrogen, **phosphate** fertilizer is still responsible for encouraging most of the world's food production. Most of the big phosphate deposits are derived from animal bones and shells. Not all sources of phosphorus are ancient, however – there is a thriving industry in mining **guano**, recent bird-droppings, for use as fertilizer.

iron oxide (hematite) is the main cement of sandstone, calcite is another chemical that sometimes takes on this role.

Other sedimentary rocks form because of dryness, not water. The **evaporites** are basically salts, formed when an area of water dries out, and later buried. The process can be seen in action today at the Gulf of Kara Bogaz in Turkmenistan, on the coast of the Caspian Sea. The Gulf is connected to the sea only by a small channel, and water evaporates faster off the Gulf than the channel can replace it, so fresh salt deposits build up. If you buy sea salt to put on your dinner, now is the time to stop. Normal salt just comes from older seas. The same process can also produce deposits of salts other than sodium chloride, such as nitrates and gypsum (calcium sulphate). If a lake dries out with no new water arriving, they precipitate out from the water in a set order reflecting their solubility.

Rocks from the fiery Earth

One important tip for would-be geologists is that you should refer to all sedimentary rocks as "soft" rocks, even if they are unbreakably hard. By contrast, igneous rocks are known in the trade as hard rocks, irrespective of their hardness. Perhaps in recognition of their deep and fiery origin, the scientists who work with them seem to have bragging rights over those who study material that might be regarded as solidified mud.

Igneous rocks form by solidifying from a liquid state. Looked at down a microscope, they are fundamentally different from sedimentary rocks. Sedimentary rocks look like more or less rounded bits of older rocks, cemented together by some other material. Sometimes the bits are very alike, as with some sandstones, where the pieces are nearly the same size, similarly sharp or rounded according to the weathering they have had, and in some cases aligned by wind or water. In other sediments, formed in more chaotic settings, there can be a wide range of grain sizes from boulders to microscopic particles, thrown together and cemented.

But in a microscope slide of an igneous rock, something else altogether is on view. Imagine heating a chunk of rock. As it warms up, the crystals with the lowest melting point will melt first. Now imagine cooling the resulting liquid. The first crystals to emerge will be those with the highest melting point. So these crystals are the largest and most perfect. Squeezed into the spaces between them are crystals of all the other minerals that make up the rock. The last to emerge may have the atomic form of a crystal but will look nothing like your mental image of one. Instead it will be forced into an arbitrary-shaped gap between the earlier arrivals.

There are two main components to an igneous rock. The most significant are the "**rock-forming minerals**". These are the ones that appear in bulk, account for most of its mass and give it its properties. The rest are bit-part players. They can have economic value, as with many metal ores, or even gold and diamonds, but are present only in small amounts.

The rock-forming minerals are mainly silicates, chemical compounds formed of silicon, oxygen and metals. On their own, oxygen and silicon add up to **quartz**. With sodium, potassium, calcium and aluminium, they yield **feldspars**, the commonest group after quartz. With magnesium and iron, the end product is **olivine**. **Micas** are minerals which form flat crystals. **Amphiboles** and **pyroxenes** are also silicate minerals, but form groups with more complex chemistry. To the expert, their composition is rich in data on the temperature and depth at which the rock formed.

Igneous rocks can be categorized along two axes. The first describes where they formed and here the principle is nursery simple. If igneous rocks form at the Earth's surface, they cool fast and have the smallest crystals. If they form deep down, they have bigger crystals because they have taken longer to cool, being surrounded by solid rock. Those that solidify near the surface or in smaller volumes cool at an intermediate rate with crystal sizes to match.

The second is the matter of how "**acid**" or "**alkaline**" the rock is. Lick a piece of lava (don't try this until it has solidified) and it will not seem either acidic or alkaline. This description really tells you how much quartz it contains. As quartz – the most common form of silicon dioxide – is the acid underlying silicates, more quartz means a more acid rock, while less means one that is more alkaline (or basic, essentially synonyms in this context). Because molten quartz is stickier than the other rock-forming minerals, having more quartz makes the rocks involved more gluey in their liquid state, as we shall see.

Starting at the top, the rocks formed at the Earth's surface are called **lavas**. They reflect the full range of chemistry within. The least acidic lavas are **basalts**, which contain little silica and are very liquid. In some parts of the Earth such as India they have formed massively thick deposits. So an alkaline volcano is the one to live next to. Better to avoid are those emitting **andesite** (a bit more acid) and **rhyolite** (the most acid). These sticky lavas produce volcanoes that erupt rarely but with extreme violence. Mount St Helens in the US is a case in point. The violence of such eruptions is enhanced if the lava contains a lot of gas, which expands as the pressure comes off it. In addition the gas can be poisonous (as with hydrogen sulphide) or might block the oxygen from reaching people and animals, as happens with carbon dioxide in incidents which have involved severe loss of life.

Once it reaches the surface, lava still has some options. It can flow along until it solidifies. Or it can be produced as ash, which can settle miles away. Rocks formed from the fusing of layers of volcanic ash are called **ignimbrites**.

In addition, lavas can be erupted under water, where they produce the distinctive shape of pillow lavas. Lavas that solidify from very gassy liquid can contain large volumes of holes and are termed **pumice**. And when lavas are cooled at very high speed, the solid that results can be a glass such as **obsidian**, not a crystal at all.

The biggest crystal sizes are encountered in rocks such as **granite**, a rock whose origin was a twentieth-century epic tale of geological science.

It obviously cools deep in the Earth, as we can tell from its large crystals, and tends to be found in the exposed roots of big, old mountain chains. But it appears in such bulk that the question that needed answering was where the rocks had gone that must have had to make room for it. In a tale told beautifully in Richard Fortey's *The Earth*, it is now known to be produced by the melting of crustal material at depth. Granite consists of at least 20 percent quartz. The most common coarse-grained rock with a less acid chemistry is called **diorite**. Again, nature laughs at these neat categories, so there are intermediates like granodiorites.

The rocks with medium grain size are far from being a point of detail. Many are formed when **magma** – liquid rock – is squeezed into lines of weakness between sedimentary rocks, where they crystallize out to form **sills** (more or less horizontal) or **dykes** (more or less vertical). Their alignment reveals the centres of action of old volcanic episodes, and they are also beautiful – see for example the Whin Sill in northern England, which has produced such lovely sites as Holy Island. These medium-grained rocks have self-explanatory titles such as microgranite or microdiorite – same chemistry as granite or diorite, smaller grain size.

The Whin Sill in Northumbria. The Romans found it a handy barrier against barbarians – Hadrian's Wall runs along the top of it (see foreground).

Rock restructuring

If a mass of molten rock is thrust into a pre-existing sediment, there is bound to be a rude thermal awakening for the soft rock concerned. The changes that ensue produce the last category of rocks we shall look at, the **metamorphics**.

Metamorphism can happen on a number of scales. If a sill or dyke is injected into a sedimentary rock, the result can be a few metres or centimetres of cooking. But when large masses of rock are dragged to the roots of a continent, cubic kilometres can be remade at a time. For example, limestone hardens to **marble** (at least to the geologist – builders and architects use the term much more loosely), while shales turn into rocks called **pelites**. If they are not too heavily metamorphosed, these can retain the layered structure they had upon their original deposition. If it is possible to split them along these planes, the rock is **slate**.

The basic ingredients of metamorphism are temperature and pressure, but the way they work is not straightforward. On their own, these two forces can turn the crystal structure of a rock formed near the surface into a higher-pressure version. But if the stress is aligned in a specific direction, the crystals will realign in sympathy. Another metamorphic effect is the separating-out of the chemical components of a rock, to give a new rock with a banded appearance.

The new minerals that form under metamorphism reveal information about the temperature and depth at which the change occurred. The temperature can be anything from about 300 to 1000°C, and the pressure can be equivalent to depths as great as 50km below the Earth's surface.

Most major types of sedimentary rock are found in metamorphic guise. The most highly metamorphosed are called **gneiss** (pronounced "nice"). These are the most banded, folded and coarse-grained of metamorphics, and it can sometimes be tough to work out what rock they are descended from. But igneous rocks are not guaranteed remission from metamorphosis. Their altered forms have names such as metagranite.

Don't forget that once rocks have been removed to these mind-numbing depths, they have to be pushed back up to the surface for us to see them. In the next chapter, on the deep Earth, we shall learn more about how this happens.

Reading history in rocks

All these rocks hold clues to the history of the Earth and, through many years of hard work by hammer-wielding geologists, this information has been painstakingly unlocked. In a saga of endeavour generally traced back to the work of William Smith in early-nineteenth-century England, the geological tale of the Earth has been assembled bed by bed. The **geological column** you see here conveys only the barest bones of their achievement. Geologists have divided up the record into thousands of units and sub-units, characterized mostly by the distinctive fossils they contain, noted the folding and faulting that distorted them, and marked both the ways they change across space and the way surprisingly similar systems can be found thousands of kilometres apart.

Although this activity is generally called geological surveying, it is very far from the flat compilation of a catalogue. Instead, it yields true Earth stories. An example of vast economic importance is the formation of coal. The rocks called coal measures are not just thick layers of coal. Instead the coal is found in seams, some many metres thick and some too pitiful to be worth mining. Between them are layers of coarse and fine sandstones, muddy rocks and even limestones. This means that the coal – laid down originally in a kind of peaty swamp – was later buried by sediments laid down in first shallow and then deep water, as finer sediments are associated with a more peaceful, deep-water, marine environment. Eventually the basin fills up and the cycle starts again with more peaty deposits.

Most geological effort has been put into describing the past 570 million years of the Earth's history, known as the **Phanerozoic** eon. This starts

Name that era...

The terms familiar from the geological column are only the most basic elements of the story. But these names themselves are worth a visit. Some are derived from the type of rock that characterizes them, as with Carboniferous for strata involving lots of coal, or Cretaceous for the chalky strata. Others come from the area where the rocks were first described, as with the Jurassic (the Jura mountains in Europe) or the Permian (Perm in Russia). Another, the Triassic, is so-called just because it divides naturally into three sub-units. Best of all for any Welsh nationalists reading this are the first three. The Cambrian is named after the Latin word for Wales, while the next two, the Ordovician and Silurian, get their names from ancient Welsh tribes.

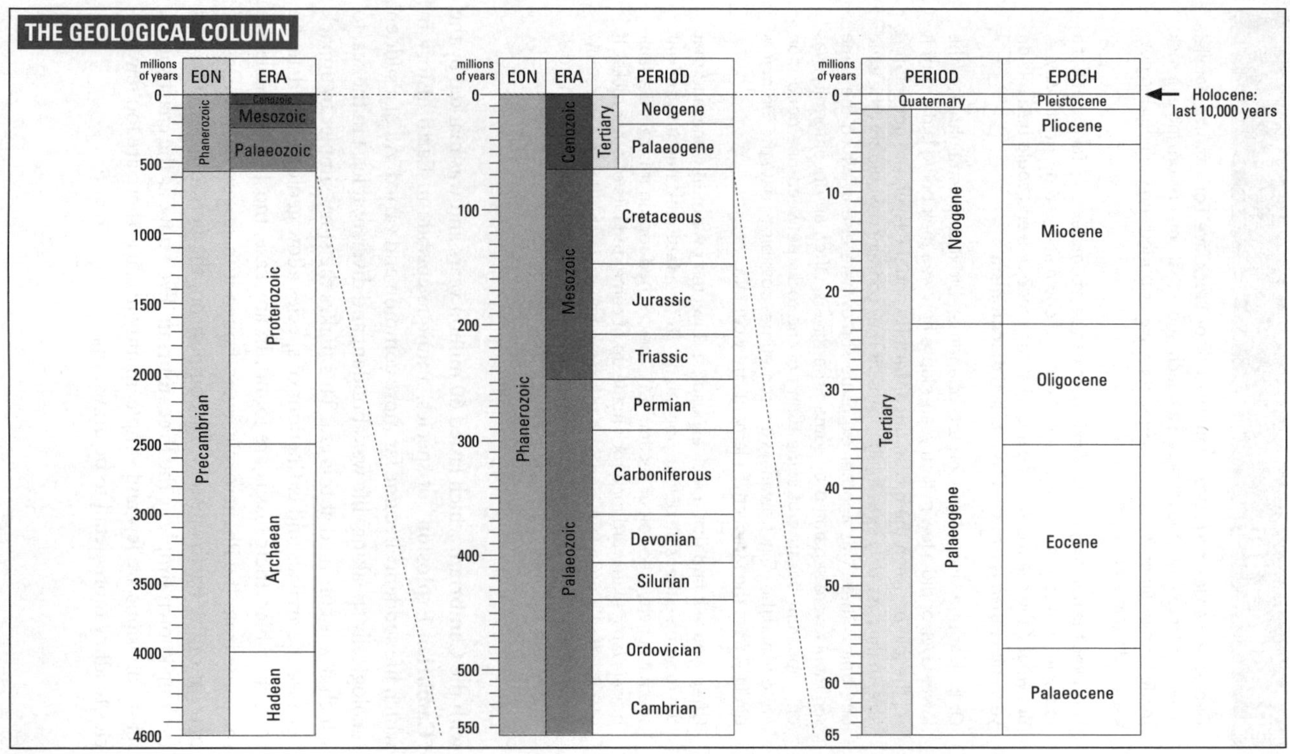
THE GEOLOGICAL COLUMN
millions of years
EON
ERA
0
500
1000
1500
2000
2500
3000
3500
4000
4600
Phanerozoic
Precambrian
Cenozoic
Mesozoic
Palaeozoic
Proterozoic
Archaean
Hadean
millions of years
EON
ERA
PERIOD
0
100
200
300
400
500
550
Phanerozoic
Cenozoic
Mesozoic
Palaeozoic
Tertiary
Neogene
Palaeogene
Cretaceous
Jurassic
Triassic
Permian
Carboniferous
Devonian
Silurian
Ordovician
Cambrian
millions of years
PERIOD
EPOCH
0
10
20
30
40
50
60
65
Quaternary
Tertiary
Neogene
Palaeogene
Pleistocene
Pliocene
Miocene
Oligocene
Eocene
Palaeocene
Holocene: last 10,000 years

Reading a bed

Geologists can learn a lot from careful study of a rock face. For instance, parallel lines, similar to high tide marks on a beach, show that you are looking at a rock made from sediments deposited in shallow water, buried and lithified, and now exposed again by erosion.

However, the full story of a rock exposure can be a complex one. Take a look at a long, uniform stretch of sandstone. It was laid down at one time, right? Maybe. But what if the sea was moving inland at the time? One end could be a million years older than the other. This is called **diachronism**.

Or look at one layer of rock on top of another. Did it come straight after the lower layer, or did a few million years elapse between their being laid down? If so, the join between them is an **unconformity** and the layers are unconformable. In the past, only examining the fossils they contained would sort the issue out, unless there was another exposure somewhere else in which the same sediments were separated by some other layers. Often an unconformity was only apparent because of some tilting of the rocks between one deposition episode and the next. Nowadays it is sometimes possible to date the sediments directly, from the radioactive elements they contain.

Often, you will find that a rock exposure is bisected by a murky line, and that the rocks on either side of it do not line up. This is a **fault**, testament to ancient seismic activity (see colour section p.4). The crushed rock in it is called a fault breccia (an Italian word for a rock made up of cemented bits of other rocks). If you can find a small fault and spot the same bed on both sides, measure how far one has moved relative to the other. Now you have established the "throw" of the fault and may regard yourself as a true geologist.

with the **Cambrian**, which lasted 60 million years and was marked by the "**Cambrian Explosion**" of species, a unique episode in Earth history in which life suddenly became far more common and varied. All of a sudden (geologically speaking) life was abundant and diverse, at least in the sea, to which it was confined at this era. This makes the story simpler to unpack because the arrivals and extinctions of species allow geologists to date the rocks in which their fossils are found. In any case, most of the rocks we see come from the Phanerozoic, even though it makes up only the last 11 percent of the Earth's history. By comparison, the period before this was for long known simply as the **Precambrian**. Even now, as the chart shows, its subdivisions are few and simple, mainly because there are too few species to allow more detail to be drawn up.

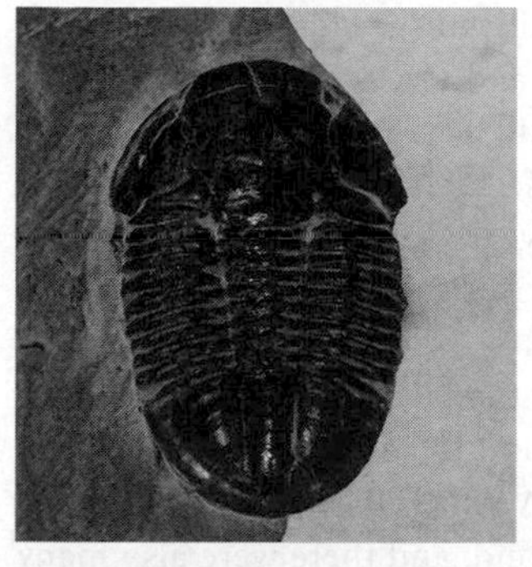

The Cambrian's most recognizable denizen is the **trilobite** (left), although, as with many of the fossils mentioned below, they lived well beyond the period which they characterized.

Most species found in the fossil record are marine ones and many are shells, often looking at first sight much like the seashells you might find today. But as ever in the fossil record, this tells you nothing about what things were like at the time. Having a hard shell increases an organism's chance of being seen in a museum in a few hundred million years. But we infer from today's ecology that there were also plenty of soft-bodied creatures around in the past. In the case of the Cambrian, we have detailed knowledge of some of them from rock formations such as the **Burgess Shale** in Canada, where freak conditions preserved a weird and wonderful array of them. It formed from material shed down a continental slope.

At this time, too, things were shifting on the tectonic front. Most of the present land mass of the southern hemisphere was joined together in one supercontinent, while most of the northern hemisphere's present-day land mass was flooded, save for a few smaller continents more or less corresponding to Siberia, northern Canada and Scandinavia today.

Although they look definite on the charts, and are often marked by substantial changes in rock type, these geological periods are human artefacts and not inherent in nature. The **Ordovician**, which ran from 510 to 439 million years ago, is a case in point. It was dreamt up to end a long-running argument about where the Cambrian ended and the next period, the Silurian, began. Its interpolation ended a lengthy period of scientific warfare among the mountains of Wales, of which the Silures and the Ordovices were inhabitants before and during the Roman era.

The most distinctive Ordovician fossils are those of **graptolites** (right), now-extinct marine organisms that floated about ancient seas in colonies up to a metre or more in length, and which appear in the rocks as sawtooth-like fossils. This period also saw the first primitive fish. Most of the land was in the southern hemisphere at this time and traces of the glaciers that formed in what is now north Africa can still be seen.

This extended glaciation of areas now far from the poles continued into the **Silurian** (439–409 million years ago), despite the drift of land, including present-day Siberia and Australia, across the Equator. Graptolites were common in this period too, and they have been used to mark out the major layers into which the Silurian is divided, which have names mainly derived from Wales and the areas of England adjacent to it such as Llandovery and Much Wenlock. The big news story in the Silurian was the arrival of plants on land, and there were also many shallow-water life forms such as corals, and freshwater animals that may have made the odd trip onshore. Comedians have often viewed this move as life's biggest error.

From Wales, the action now moves to south-west England, as the next period, the **Devonian** (409 to 363 million years ago), takes its name from the county of Devon. The massive sandstone deposits of that era in Devon are known as the Old Red Sandstone. They were laid down by erosion from the Alpine-size mountain range which dominated Wales at that time but which has now been worn away to nearly nothing. During the Devonian, Europe and North America were equatorial territory. Although life expanded, with many more species of fish and the growth of forests and insects on land, the end of the Devonian was marked by a mass extinction of many marine species. Its cause is still a matter of speculation.

Most of these geological periods are classified in the same way the world over but the **Carboniferous**, which lasted from 363 to 290 million years ago, is an exception. Although North Americans do use the term, they prefer to think of these rocks in two halves, the Mississippian and the Pennsylvanian. But the word "Carboniferous" is an eloquent one. Although there are some important younger coal deposits, the Carboniferous ones are the biggest and the best. There are no older ones because before this time, there were too few land plants to rot down and make coal.

The North American distinction between the Mississippian and Pennsylvanian is more or less reflected in Europe by a division of the Carboniferous into the "Lower" and "Upper" (see box opposite). The Lower Carboniferous rocks tend to have accumulated in more or less shallow seas as the continents converged, while the Upper Carboniferous, as we have seen, produced the coal measures through the steady, cyclical

Superposition

The Lower and Upper Carboniferous owe their names to the deepest theoretical tenet of geology – the Principle of Superposition. This states that the new stuff is on top of the old stuff. It is credited to Niels Stensen (a Dane, who like many intellectuals of the day usually went by a Latinized version of his name, in his case Nicolaus Steno) in the mid-seventeenth century. Sometimes, because the Earth is active and throws material about by faulting and folding, the older rocks end up on top of the younger ones. But the geologist would still call the older ones "lower" and the newer ones "upper".

drying-out and inundation of coastal areas. The lush forests whose plants were later to turn into coal were home to the biggest animals yet seen, with large amphibians and reptiles on the land and in the sea.

However, these creatures' descendants did not enjoy what happened next, in the **Permian**, which ran from 290 to 248 million years ago. It was at this time that Wegener's **Pangea** was formed, with essentially all the land we see today gathered in a ring around a sea called Tethys, while a single mighty ocean, Panthalassa, took up most of the Earth's surface. The Permian was marked by many mass extinctions that removed almost all land and marine species. The lower Permian has many signs of life such as limestones, rocks which are essentially remains of shells and corals, and the coal accumulation of the Carboniferous continued. But the upper Permian is characterized by rocks which indicate drying out and cooling. The cause may have been a reverse greenhouse effect generated by a lack of carbon dioxide in the atmosphere after so much had been swallowed up into coal measures.

The end of the Permian is also the end of the **Palaeozoic** era, which began with the Cambrian. These longer pieces of geological time are not vital to understanding but they mark big changes to life on Earth through time.

The next of these eras, the **Mesozoic**, starts with the **Trias** (or the Triassic), which is so continuous with the Permian that they are sometimes lumped together as the Permo-Trias. It lasted from 248 to 208 million years ago, during which time the map of the world did not change much. Volcanic lavas, and sedimentary rocks of both land and marine origin, formed in and around Pangea. Many of the land-based ones are associated with dry or even desert conditions, such as salts from the drying out of lakes, and desert sands. The ancestors of both dinosaurs and warm-blooded present-day mammals gained ground at this time.

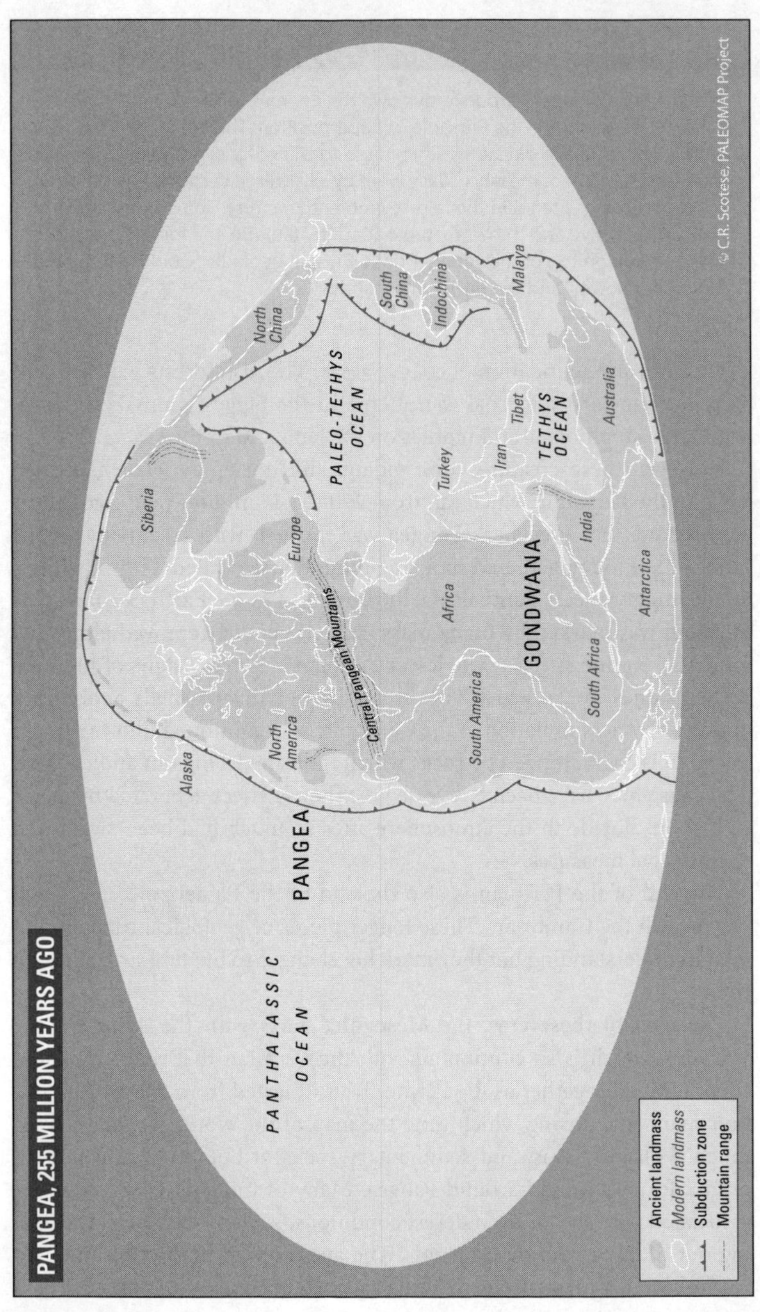
PANGEA, 255 MILLION YEARS AGO
PANTHALASSIC OCEAN
PANGEA
PALEO-TETHYS OCEAN
TETHYS OCEAN
GONDWANA
Alaska
North America
Central Pangean Mountains
South America
Europe
Siberia
Africa
South Africa
Antarctica
India
Turkey
Iran
Tibet
Australia
North China
South China
Indochina
Malaya
Ancient landmass
Modern landmass
Subduction zone
Mountain range
© C.R. Scotese, PALEOMAP Project

Next, from 208 to 146 million years ago, comes the only geological period with its own series of Hollywood blockbusters – the **Jurassic**, named after the Jura mountains of France and Switzerland.

The Jurassic rocks of Europe are probably the easiest place on Earth to find fossils. Mostly they are shells and corals formed in warm water, but the origins of big-league fossil collecting are in these rocks too, along the Jurassic coast of England. These rocks yielded the fossils that led to the coining of the term dinosaur. The first bird, archaeopteryx, also dates from this era. Most of these fossils are found in rocks that developed in shallow water. But the Jurassic saw the final break-up of Pangea as a result of sea-level rise and renewed plate tectonics. Much previous land was flooded, and North and South America were separated by a seaway. India became an island, as did Antarctica, one of the few land masses to be more or less in its present-day position in this era, but at that time attached to Australia.

The final period of the Mesozoic is also the point at which the map of the world starts to look more recognizable, with the Atlantic starting to open and Australia heading north. Called the **Cretaceous** (creta is the Latin for chalk), it lasted from 146 to 65 million years ago. The high sea levels of this period left vast chalk deposits across large areas of the world, with the white cliffs of Dover the most sung-about. The chalk is formed from coccoliths, the shells of tiny animals. At this time there was also extreme tectonic action, with huge volcanic outpourings forming thick deposits of basalt called the Deccan Traps, in India.

The Seven Sisters cliffs in Sussex, England – made of fossil sea creatures and today a prime fossil-hunting spot

The end of the Cretaceous is probably the most discussed event in Earth history (see p.54), and will remain a talking point. However, the asteroid impact idea would now take some dislodging from both science and the popular imagination.

The disappearance of the dinosaurs at that time meant that our mammal ancestors dominate the rest of history – or at least, they make up most of the big land animals. Although whales are the largest animals, fish still dominate the sea while the only notable flying mammals are bats.

The rest of the geological column is the **Cenozoic** era, and this is the one we are living in. Its rocks are much-studied and their divisions and subdivisions have been rejigged with some enthusiasm over time. Nowadays its lower half, the **Tertiary**, is itself divided into the Palaeogene and the Neogene. The **Palaeogene** lasted from 65 to 23 million years ago and was marked by further opening of the Atlantic and the beginning of the formation of the Himalayas as India pushed north into Asia and the crust between the two became ever more compressed. This process continues today. Major evolution on land included the development of ever larger mammals.

But quality counts as well as quantity. The **Neogene**, from 23 to about 1.8 million years ago, is notable for one big item in evolution, the separation about 5 million years ago of human ancestors from their last ape relations. It is divided into the older, and still more familiar, **Miocene** and **Pliocene**. The boundary between these is best left to the professionals. It is defined by the percentage of still-existing molluscs in fossils at different times.

The final phase of the story, the **Quaternary**, lasts just 1.8 million years. In the geological nomenclature it has been renamed the **Pleistogene** and has equal status with, say, the Jurassic, which lasted 70 million years. But it is only reasonable that we care more about our own times, especially as there is more detailed information available on them.

All of this period except the last 10,000 years is called the **Pleistocene**. This is too short a time for much to have happened on the plate tectonics front. Instead it was marked by the formation of most of the landscapes of the northern hemisphere through a number of episodes of more or less severe glaciation. Go for a country walk almost anywhere in the northern hemisphere and you will see the traces, described on pp.219–23.

All this cold did not prevent *Homo sapiens* spreading out from Africa across the world. Indeed, the lower sea levels caused by water that is today in the oceans being turned into ice made the task simpler, for example providing a land bridge between what are now Siberia and Alaska (see p.226).

The time period we are living in is called the **Holocene**. This has been persistently warmer than an ice age and is referred to as an interglacial because the pessimists in charge of the terminology think that the warm period is merely a prelude to the next ice age. As well as the gross effects, it is apparent that the growth and reduction in ice volumes in polar and high mountain regions has other more subtle consequences, affecting climate and temperature far from the ice itself.

However, it is also fair to say that the increasing numbers and impact of humans on the world have been among the most striking changes of the Holocene era. For more on our impact on the world, turn to Chapter 8.

4

The deep Earth

The deep Earth

Too few people read Jules Verne these days. But if you have read *Journey To The Centre Of The Earth*, or seen the film, do your best to forget it. Engineering knows no material strong enough to make caves to lead you to the centre of the planet. If any did open up for some reason, they would collapse again long before James Mason and his companions could make use of them.

Nor is pressure the only problem with day trips to the deep Earth. Anyone going to the centre of the planet would have the weight of the world on their shoulders, but they would also be disagreeably warm. Mines scratched just a few kilometres into the Earth's crust already need substantial air-cooling equipment to allow people to work in them. As you get deeper, the heat carries on building.

The sheer difficulty of examining the deep interior of the Earth means that while spacecraft have made objects billions of kilometres across the solar system familiar to us, we have yet to make any significant incursions into the ground below our feet.

Instead, the deep Earth is a showcase for what science can find out about something it can never see. And this is not mere scientific virtuosity for the sake of it. There are plenty of things going on thousands of kilometres below our feet that influence our everyday lives here at the Earth's surface.

But what do we mean by the deep Earth? Perhaps the easiest way to define it is to regard it as the Earth below the crust. That makes more sense than adopting some arbitrary depth as the boundary. For the tidy-minded it has the disadvantage – as you remember from Chapter 3 – that the crust below the continents is a lot thicker than below the oceans. So the deep Earth in mid-ocean could start 10km below sea level, but in a hefty mountain range it could be 80km down.

However, nobody can argue that either of these is exactly shallow by human standards. The deepest mine in the world, East Rand gold mine in South Africa, extends to a mere 3585m below the surface. At the time of writing, the deepest well ever drilled is 12,262m, in the Kola Peninsula in Russia (see box on pp.104–05).

This is impressive technology, but the radius of the Earth is 6371km. So even the most ambitious digging and drilling have yet to make it even 0.2 percent of the way to the centre. Another way to think of it is that the mantle makes up 84 percent of the solid Earth's volume, and the core another 15 percent, but our direct exploration is confined almost entirely to the outer 1 percent that we call the crust.

Probing the inner depths

We may not be able to see it first hand, but we have not let that stop us finding out about the deep Earth. There is one tool above all that helps us to do so: the **earthquake**. As we saw in Chapter 3, the shock waves emitted by earthquakes have been used to detect the Mohorovičić Discontinuity at the base of the crust. But they can tell us much more than this, and allow us to see all the way to the Earth's centre.

Earthquake **shock waves** come in two forms. **P waves** are "compression" waves in which the matter making up the Earth is compressed and then pulled apart by the passing wave, in the direction of its travel. With **S waves**, or shear waves, the material moves from side to side at right angles to the direction of the wave. Sound in air is a P wave. Watch the sea and you are viewing S waves.

A sizeable earthquake will produce waves of both types detectable across most of the Earth. They are measured and timed by a network of seismometers across the planet, mostly in universities and government laboratories, which have now accumulated many decades of data.

S waves move more slowly than P waves. More intriguingly, they cannot travel in fluids. As we shall see, the Earth's core is liquid, consisting mainly of molten iron. Its existence was discovered back in 1906 because for any earthquake, there is a zone on the Earth's surface at which the resulting S waves cannot be detected. This happens because these waves would need to pass through the core to get there. Called the **Shadow Zone**, this area starts 105° from the epicentre of the earthquake. It took seismology a little longer to prove the existence of the inner core. But by 1936 more detailed analysis of the time it took P waves to travel through the core had shown that there

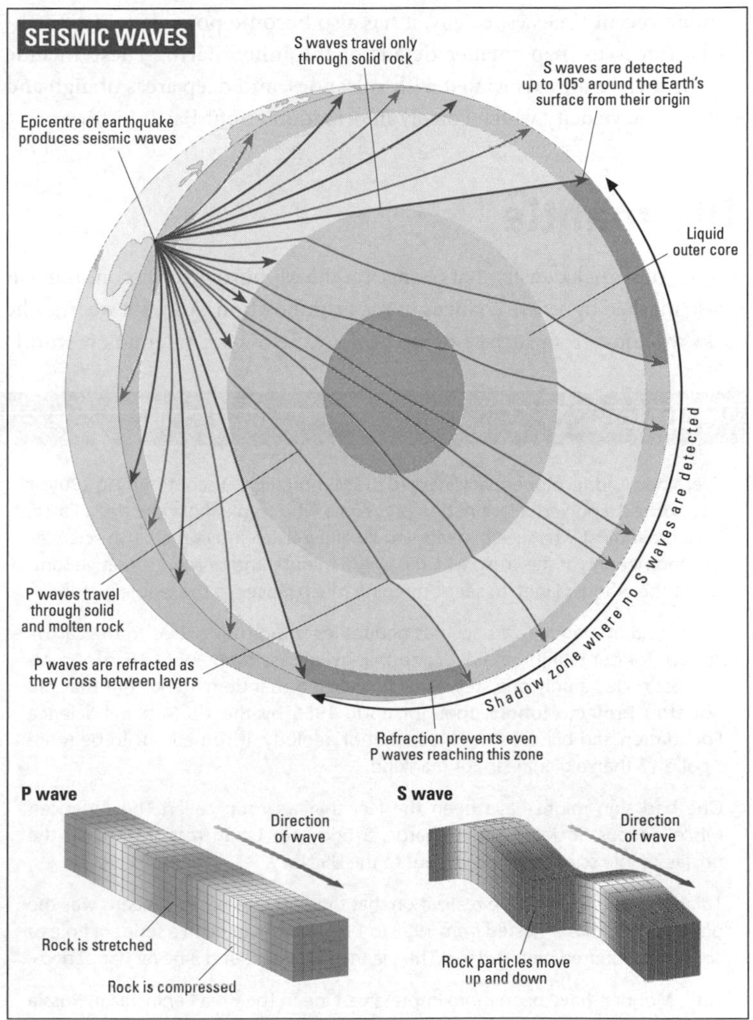

was an area at its centre through which waves were travelling faster than they would through a liquid. This showed that the core must have a solid centre.

Seismometers can detect a rich range of earthquake-wave effects. As we saw in Chapter 3, waves can be reflected off the Moho, or can be refracted along it. The same is true for the boundary between the mantle and the core. For any earthquake, detailed maps of the routes taken by seismic waves can be drawn up. Over time, this information has allowed the depths of major structures within the Earth to be mapped in detail.

In more recent times especially, it has also become possible to use earthquake traces to map smaller details of the inner Earth. These include molten rock bodies associated with volcanoes, and deep areas of high and low seismic velocity which convey information about Earth history.

The mantle

It is not only rock density that changes at the Mohorovičić Discontinuity. It is also marked by severe changes in rock composition. Above it we find the rocks we know from surface-based geology. Below, we enter a new world.

Moholes

Ever since Andrija Mohorovičić named the Mohorovičić Discontinuity in 1909, it has exerted a powerful fascination on people who think about the deep Earth. Taking a look at what lies beneath would allow us to move the Earth sciences beyond peering at the crust, which is only a minute and unrepresentative sample of the whole planet, to see something much closer to the whole picture.

It is true that mantle rocks such as peridotites bring themselves to the Earth's surface for our delight unaided, and that even basalt, one of the Earth's commonest rocks, is only one step away from being mantle material. But that did not stop **Project Mohole**, thought up in 1956 by the US National Science Foundation and billed as the Moonshot of geology. It turned out to be more Apollo 13 than a giant leap for mankind.

One bad sign might have been the fact that a group called The American Miscellaneous Society was in charge, a body set up in response to all the unclassifiable science proposals put to the US Navy.

Things started OK with the realization that the thin crust of the oceans was the place to drill. Phase 1 lasted from 1958 to 1966, and resulted in a series of holes a couple of hundred metres deep. The planned Phases 2 and 3 never happened.

Later Moholes have been more impressive. One in the Kola Peninsula in Russia ran to over 12km, but still did not get through the crust. It was about 2km deeper than the deepest well so far drilled for commercial purposes, a gas well in Oklahoma.

While the world seems to have moved on from the concept of Mohole drilling, there is still plenty to be learned from deep holes in the Earth. In Germany, a 9km hole has been drilled to find out more about the way fluids move underground and how fault systems develop. It turns out to be surprisingly easy to simulate a swarm of earthquakes by building up pressure in the well.

The boundary between the crust and the mantle below is no hermetic seal, however. It is more like something out of biology, maybe the stomach wall, an important but permeable boundary across which there is lively trade. The Earth's crust and mantle are constantly interacting, and we have the rocks to prove it. In numerous parts of the Earth's surface, volcanoes bring small pieces of the mantle to the surface. At mid-ocean ridges, much larger volumes of mantle rock can appear.

But there are limits to how much these rocks can tell us about the deep Earth. As material from the mantle rises towards the surface, it melts because the pressure drops. But it does not do so uniformly. The minerals with the lowest melting point go liquid first and those that melt at higher

In Iceland, much shallower wells drilled in a volcanically active area are also showing research promise. The **Iceland Deep Drilling Project** exists to extract heat from Iceland's geothermal reserves at a far higher temperature than the steam and hot water that emerge at the surface. The ambition is to drill down to about 5km where there will be superheated steam on tap for electricity generation and other uses.

Finally, 23 organisations from around the world are so convinced about the wisdom of deep drilling that they fund the **Integrated Ocean Drilling Program**, under which a drilling rig called the Joides Resolution sails every bit of blue on the map to answer questions about the make-up of the Earth's crust by drilling into it.

Indeed, in April 2005 it turned out that the crew of the Resolution might be about to deliver us a Mohole by stealth. They had already reached the base of the crust and might break through to the mantle some year soon.

But have no fear. As we have seen, the mantle is liquid enough to creep along at a rate of a few centimetres a year, pulling the continents along and driving plate tectonics. But it is far too viscous to spurt out of a possible Mohole, volcano-style, especially as the hole would be well under a metre in diameter. Instead, it would be a feat of high technology to get rock samples from the top of the mantle back to the surface to go under geologists' microscopes.

For further information see:

International Continental Scientific Drilling Program www.icdp-online.de

Ocean Drilling Program www.odp-tamu.edu

Project Mohole History www.nas.edu/history/mohole

temperatures may remain solid throughout the process. The upshot is that the material that makes its way upwards to appear at the mid-ocean ridge volcanoes as liquid basalt contains far more quartz – is far more acid, in petrologist-speak – than the mantle itself. By contrast, the leftover rocks – one is called Harzburgite – are silica-poor (more "basic" or "alkaline").

The upper half of the mantle consists mainly of a rock called **peridotite**, which itself consists mainly of the mineral olivine, a silicate of iron and magnesium which is denser than the crust rocks found above.

As the melting rock that finds its way to the surface shows, the crust and mantle are really a single system. To be exact, it is the crust and the upper half of the mantle that work together to drive the plate tectonic machine, and they are referred to collectively as the **lithosphere**.

Boiling toffee

The mantle is not a solid structure. Instead it is fluid, although its viscosity is so extreme that a piece of it would look indistinguishable from solid rock. And it is constantly in (admittedly slow) motion.

You may remember from school that there are three ways for heat to get from place to place. One is **radiation** – basically when it moves through more or less empty space. Another is **conduction**, where it moves through solid matter. And the third is **convection**, where it is transferred by the movement of warm matter. There will be plenty of convection in future chapters of this book, because it is central to the running of the atmosphere and the oceans, but it is also the driving force behind the movement of the mantle.

Think of a pan of boiling toffee. As it boils, hot material moves up to the surface where it begins to cool. After it has cooled, it is dragged down again. The structures this process forms are called **convection cells**. The lithosphere works in pretty much the same way. Zones such as the Mid-Atlantic Ridge where new crustal material is being created mark the upswellings, while areas where crust is being removed, such as the deep Pacific off the coast of South America, mark the downward edges. Thus, the map of the Earth's seismic zones on p.67 serves as a surrogate map of the upper mantle.

Even within the cells, however, the upper mantle is an active and far from uniform place. One study which examined seismic waves passing through the region below the south-west United States showed that there were large volumes of the upper mantle with unexpectedly low seismic velocities, indicating that they consisted of molten material. The overall

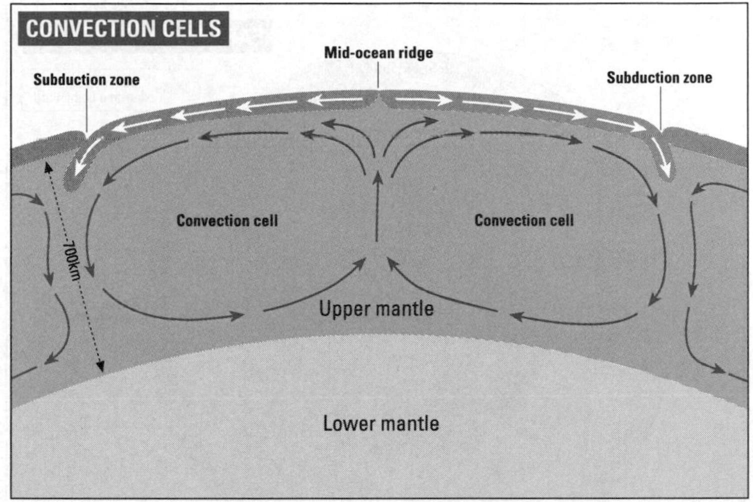

picture it revealed was of upswelling and subduction on a smaller scale than the continent-sized movement we normally think of in the context of plate tectonics. The topmost 400km of the lithosphere seems to have a fine structure of smaller convection cells, most readily detected near active ocean ridges, which act as the final, finer stages of the big plate tectonic machine.

Research carried out in Russia and elsewhere suggests that this convection may have significant effects by concentrating metals and minerals. Some big Russian mineral deposits, such as the nickel deposits at Norilsk, are associated with places where three convection cells meet, or did in the past. The scientists involved say that this method may allow the location of giant mineral deposits to be predicted.

The toffee model may sound crude, but in the hands of laboratory scientists it can produce genuine insights. For instance, you can see how big the cells are on top, but how deep are they? The rule seems to be that they tend to be about as deep as they are wide, and the same rule applies to the Earth itself.

The mantle runs down to about 2700km below the Earth's surface, but if you look at a map of the tectonic convection zones that make up its surface, you will see that they tend to be roughly 700km across. So the lithosphere should be about 700km deep. Since the crust is on average 30km deep, this means that the upper mantle is about 670km deep on average.

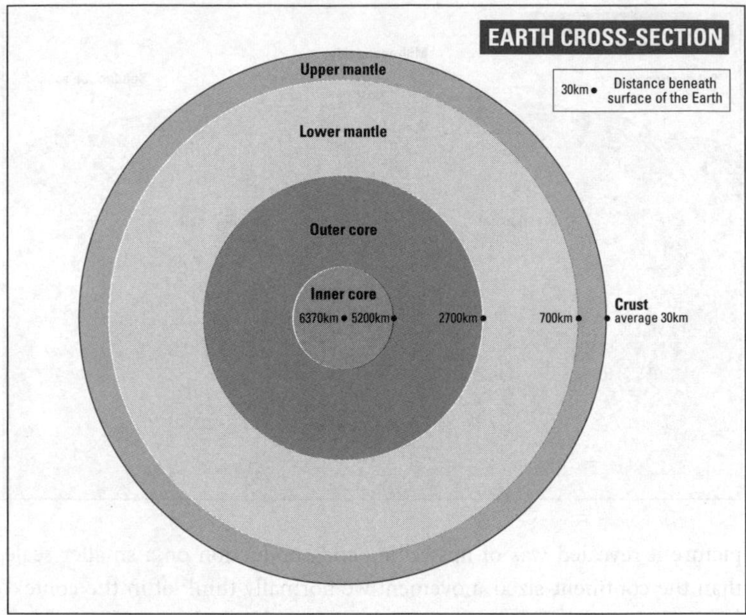

The lower part of the upper mantle is more liquid than the upper portion because the immense pressures break down the structure of some of the minerals that make it up, but at about 700km below the surface there is evidence of a complete change of composition.

The signs are that the lower mantle is made up of rocks richer in silica and poorer in iron and magnesium than the upper mantle, in other words that it is more acidic. But while we can look at the peridotite of the upper mantle and the basalts derived from it, our knowledge of the lower mantle is indirect and likely to remain so.

However, we do know some things about even this remote layer. One is that it is convecting just as the upper mantle does. This is a mathematical necessity because the heat produced in the deep Earth has to go somewhere. The lower mantle runs from about 700 to 2900km down and seems to be made up of a single layer of convection cells.

The spookiest idea in this book

Just as with the crust and the upper mantle, the Earth's upper and lower mantle seem not to be separate layers at whose boundary nothing but heat is transferred. At the very base of the lower mantle, seismic studies have

Diamonds

Thanks to the rise of feminism, it has ceased to be acceptable to regard them as a girl's best friend. So what are diamonds? Everyone knows that they are a form of carbon, but just how they got their distinctive hard and dense form is a story that tells us a lot about the deep Earth.

There are two main forms of carbon, **diamond** and **graphite**, along with some rarer types including the Fullerenes. In graphite, the atoms are arranged in hexagons making up flat sheets. That is why it comes off when you rub it on paper. Mixed with clay and other ingredients, it is the "lead" used in pencils, despite not containing any lead. By contrast, diamond consists of carbon atoms in a three-dimensional structure that they only adopt under extreme pressure. Because every atom is bonded to four others, the result is a very hard crystal.

The artificial diamonds used in drilling, and in jewellery for someone you are not terribly keen on, are made by squeezing carbon in a mechanical press until it changes phase to the more compact diamond form. Diamonds have a density of about 3.5 grams per cubic centimetre, compared to graphite's 2.3, so applying pressure is bound to make graphite turn to diamond at some point.

To achieve this effect inside the Earth, you need only take graphite to a depth of about 120km. There the pressure is about 40,000 times that of the Earth's atmosphere at sea level and the temperature is up to about 900°C, which also speeds up the process. It appears from analysis of the impurities they contain that diamonds can form from rocks that are common in the Earth's mantle such as peridotite. They also tend to be very ancient, between 1 and 3.3 billion years old.

But once a diamond has been made it has to get to somewhere near the Earth's surface for us to find it. This involves one of the most violent events in nature, a type of volcanism which has been seen at its most abundant in South Africa and is called a **Kimberlite** volcano, after Kimberley in South Africa. The volcanic material cuts its way up through the Earth's mantle and crust at a pace which finally becomes near-supersonic. No Kimberlite eruption has ever been seen; if one did occur it would be a major natural disaster. For blushing brides, however, they are absolutely essential. A diamond brought slowly to the surface from the mantle would gradually revert to graphite on the way up. But bringing them to the surface at this speed freezes them in diamond form. Although mineralogists point out that all the world's diamonds are gradually reverting to graphite, the timescale involved is billions of years.

Even in Kimberlites, diamonds are rare and thousands of tonnes of rock are shifted to find a few kilograms of them. They are also extracted from rivers that erode Kimberlites.

Very small diamonds can also be created by other high-pressure events in nature. They are sometimes formed during metamorphism inside the Earth, and they are found in meteorites as evidence of shock waves in deep space.

For more information, see brysonburke.com/aboutdiamonds.html.

detected a layer with distinctly different seismic properties from the rest of the mantle.

Known as the **D" (D prime prime) layer**, this seems to be a mass of big slabs of matter hundreds of kilometres across. One of the most challenging ideas in the Earth sciences is that this layer may be a graveyard (the term has been used by other writers but there is no real way round it) of old continental crust which has been subducted and, instead of melting and being recycled as classic plate tectonics would suggest, has slithered ever lower until it hit the base of the mantle. The idea of those dead ex-continents trapped for ever thousands of kilometres below our feet is almost enough to persuade Hollywood to make *Journey To The Centre Of The Earth II*.

But the idea that fossil continents are lurking far below our feet is only one possible explanation for the seismic velocity changes seen there. The material seen at this depth could have welled up from the core, or it could be old ocean crust, rather than continental crust. Alternatively, it could be the product of some mysterious chemical reaction between metal and silicates. Certainly the layer varies hugely in thickness, from hundreds of kilometres thick to little or nothing. It may be that, over time, movement in the crust and mantle piles up the bits of continent in some places in preference to others. While some continental rocks would melt at this depth, the immense pressure means that some would not.

It is hard to imagine the probe that would allow us to prove empirically that D" is indeed a continent graveyard. But it is becoming possible to see the deep Earth in more detail. It may become feasible to image the mantle in a way that will allow smaller convection cells to be viewed as well as the gross ones. In the same way, we now speak about what the lower and upper mantle are "made of", but it may turn out that their composition varies from place to place in ways we can map.

Making volcanoes

If all this seems a long way away from everyday life, think again. Richard Muller of the University of California at Berkeley points out that shifting material at the core–mantle boundary (the CMB, if you are meeting geophysicists you need to impress) could have effects at the Earth's surface. There could be landslides there which might encourage volcanism, arguably the most spectacular of natural phenomena and definitely one of the deadliest.

Almost by definition, a volcano involves molten rock coming up from inside the Earth. This means that the place you are least likely to find one is somewhere where the Earth's crust is at its thickest. So the Himalayas, with many tens of kilometres of mountain root below the surface, are no place to look for a volcano. Mostly, as we have seen, they are found where the crust is thin and energy and material are rising from the mantle, or where material is being subducted back into it.

So far, so simple. In both of these cases, we can see just why there would be a volcano in a specific place, whether it is in the Andes – behind which rock is being subducted – or along the Mid-Atlantic Ridge, where it is being brought from below.

But glance at the map of the Earth's volcanoes on p.67 and you will see that things are not that simple. There are some volcanoes that just appear where they are not meant to. If you are puzzled, you are in good company because the best minds in geology and geophysics have been baffled as well. It is not as though the volcanoes in question are a few oddities that can be quietly ignored. They include one entire state of the US – **Hawaii** – which is the biggest active volcanic system on Earth. But tectonically, Hawaii is also in the middle of the Pacific plate, where no volcano has the right to be.

The plume theory

The classic explanation for these volcanoes is that they are produced by **hot spots** in the mantle. These melt the rock above them and burst out onto the Earth's surface. And it happens on a grand scale. From the bottom of the Pacific near Hawaii to the peak of Mauna Kea, the highest point on the islands, is 10,200m of vertical climb.

This version of events is supported by determining the ages of rocks in the Hawaiian chain. The idea is that there is a "**plume**" of heat rising from the mantle. It stays still, but the slow movement of the Pacific plate pushes ocean crust over it through time. The result is a chain of islands getting steadily younger from northwest to southeast. Today, active volcanism is seen only on the south-easternmost of the islands, Big Island, but may revive on the next-oldest, Maui. To the northwest the islands get steadily older, with Kaui, last in the chain, 5 million years old. Beyond it come a string of seamounts, islands which once poked above the ocean

Live volcanoes

Nobody knows how many active volcanoes there are on the Earth. For one thing, activity does not necessarily mean that a volcano is spurting lava right now. Signs of life in recent decades that might be resumed in the future are enough for a volcano to be viewed as active. In August 2005, some thorough people at Michigan Technological University in the US had a stab at cataloguing the Earth's active volcanoes (see www.geo.mtu.edu/volcanoes/world.html). They found:

- **One in the Antarctic**, Mount Erebus
- **Two in mainland Europe**, Stromboli and Etna, both in Italy
- **Six in Iceland**
- **One in mainland South America**, plus one on the Galapagos Islands in the Pacific
- **One in the West Indies**, on Montserrat
- **Sixteen in Central America**, including Popacatapetl in Mexico, five in Guatemala, two in Costa Rica, seven in El Salvador and one in Nicaragua
- **Three in the continental US**, including Mount St Helens
- **Five in and around Alaska**
- **Three in Hawaii and two others in the Pacific**
- **Two in New Zealand**
- **One in Japan**, Mount Unzen
- **Four in Kamchatka, Russia**
- **Ten in the Indian Ocean, Indonesia and the Philippines**, including Mount Pinatubo, the site of spectacular volcanism in the 1990s
- **One in Africa**, on the Cape Verde Islands

All of these are active or have been recently. To them must be added the large number of ocean ridges where more or less continuous volcanism is in progress.

Every Wednesday, the US Geological Survey and the Smithsonian Institution issue their weekly alert on volcanic activity. It excludes long-running activity and sea-floor volcanism, focusing on new and potentially hazardous volcanism. See www.volcano.si.edu/reports/usgs.

For more general information see:

Cascades Volcano Observatory vulcan.wr.usgs.gov/home/html

USGS Volcano Hazards Program volcanoes.usgs.gov

and have now been eroded away. They are known as the Emperor Chain and are mainly named after past emperors of Japan. To the southwest of Big Island is an active volcanic zone of sea floor, Loihi, which promises in a few thousand years to form the next Hawaiian island.

Who could possibly disagree? Plenty of people in recent years. The trouble started when it became possible to make seismic images of the deep Earth below possible plume sites such as Hawaii and Iceland. The image of a plume cutting through mantle and crust like a welder's torch is appealingly simple, but it proved hard to detect plumes running to anything like the depth that the theory called for. They ought to start not far short of the Earth's core but sceptics found them elusive, although plume supporters claimed to observe them in numbers.

Instead, doubters point to experimental work that suggests that continental plates may be less indivisible than had been supposed. They may crack open to allow material from the top of the mantle to reach the surface. This might explain volcanoes far from plate boundaries, such as Hawaii and Yellowstone in the US. It is also a possible explanation for **Iceland**, which, although it is at an active centre for ocean floor production, generates far more lava than anywhere else on the Mid-Atlantic Ridge.

Once you accept that the mantle is a complex place where subduction and other effects might produce large volumes of rock with low melting points – potential future volcanoes – it may not be necessary to have plumes from the very deep Earth as well. Either way, the plume argument shows how something in science that everyone thought was obviously true can become disputed overnight and may turn out to be false, or only part of the story.

Types of volcanoes

Whatever causes them, volcanoes are objects of awe and fear for humans, and with good reason. They come in several main types.

Those with the runniest lavas are least dangerous, as we have seen (see p.85), because they produce the least violent eruptions. But they also cover the most territory. The lava they produce forms huge **shield volcanoes** like the islands of Hawaii, or builds up into massive areas like the Deccan Traps of India or the Colorado River plateau in the US. These are so unlike our conception of a typical volcano that they tend to be called **Large Igneous Provinces** rather than volcanoes.

Shield volcano: the massive Mauna Loa on Hawaii is the Earth's biggest

When things get a little stickier, the result is a classic volcanic shape like the one a child would draw, exemplified by Mount Fuji in Japan, last active in 1708. These very beautiful mountains are subject to sudden change when they do erupt, as the removal of a large piece of Mount St Helens showed in 1980. They are called **stratovolcanoes** – layered volcanoes – and consist of mixed layers of ash, lava and other components.

Mount Fuji: national symbol of Japan and the definitive volcano shape

Other types of volcanism can produce cones that consist almost entirely of ash, called **cinder cones**. They are produced by gassy lavas and are too weak to sustain a big lava lake like a classic volcano, so lava often runs out of the side instead. The best known, Paricutin, just appeared in a field in Mexico one day in 1943 and by the time it had finished, 25 square kilometres of land had a new lava coat.

Paricutin: appeared in Mexico in 1943

The least familiar volcanoes are those that appear under the sea, typically at centres of ocean floor spreading. As well as fresh lava, these volcanoes, known as hydrothermal vents, produce a huge amount of hot or even boiling water, depending on their depth below the surface, and some are called "**black smokers**" in honour of their carbon-rich dark emissions (see colour section p.5). The energy released by these vents doesn't go to waste, as they are home to a lively ecology of fish, shellfish and other animals. Some live in ways that would not be possible elsewhere on Earth, such as bacteria which depend on sulphur emitted from the vents.

Stand clear

Volcanoes are beautiful but they are also hazardous. When Mount St Helens erupted in 1980, the wave of debris, steam and gas that surged out moved at up to 110kph, and 540 million tonnes of ash were emitted.

Over the centuries, volcanoes have been responsible for many thousands of deaths (see box on pp.116–17). There are many ways a volcano can kill. For example, when Tambora in Indonesia erupted in 1815 it killed 12,000 people directly, but 80,000 more died subsequently from starvation.

The most direct hazard from any volcanic eruption is being engulfed in a cloud of ash and air running downhill from the volcano. Known as **nuées ardentes**, these burning clouds can travel at speeds far too fast to escape. In 1902, a nuée ardente killed the entire population of St Pierre in Martinique – apart from one man in the safety of the condemned cell in the prison (see www.mount-pelee.com). These eruptions are called Plinian in honour of Pliny the Younger's description of the destruction of Pompeii

Killer volcanoes

It is rare for volcanoes to kill in numbers comparable to the death tolls associated with large earthquakes. However, the US Geological Survey lists 35 volcano eruptions since 1500 that are recorded as killing over 300 people each. The list is probably not perfect, because for much of history, birth registers, electoral registers and the like were unknown or imperfect.

In 1996 and 1997, the **Soufrière Hills** volcano on the small West Indian island of Montserrat killed twenty people, buried the island's capital under 12m of mud, and forced two-thirds of the population to flee abroad. The damage occurred even though the volcanism had been preceded by known precursors of a major eruption such as swarms of earthquakes, mud flows, ash eruptions and dome-swellings.

In 1991 the eruption of **Mount Pinatubo** in the Philippines caused over 800 deaths and widespread destruction as it covered over 100 square kilometres of Luzon Island in volcanic ash. This eruption produced a new 2.5km caldera (a wide volcanic crater) at the top of the volcano, which lost 300m in altitude.

About 23,000 people were killed in the 1985 eruption of **Nevada del Ruiz** in Colombia. This volcano has been active for many centuries and its very acidic lava means big explosions. Most of the casualties were in the town of Armero, which was overwhelmed by a lahar, a mudflow of volcanic material (see p.118).

Most extraordinary were the events of May 1902 in the West Indies. On the 7th, the **Soufrière** volcano in St Vincent erupted, killing nearly 1,700 people, and a day later **Mount Pelée** erupted in Martinique, destroying the city of Saint-Pierre and killing perhaps 28,000 people. The type of eruption involved, with very damaging red-hot avalanches called nuées ardentes, is called Pelean as a result of this disaster.

Further back in history, the eruption of **Krakatau** (Krakatoa) in Indonesia in 1883 (see p.118) may have killed about 36,000 people and is thought to have produced waves detectable in Europe. Ash fell over 1000km away and 40m-high waves devastated some 165 coastal villages. Barometers all over the world detected the pressure changes from the eruption. Admirers of J.M.W. Turner's sunsets may be interested to know that their peculiar intensity could well be a reflection of increased levels of dust in the atmosphere after the eruption. This may also explain the blood-red sky in Edvard Munch's *The Scream*.

The 1815 eruption of **Tambora** in Indonesia is probably the most deadly ever, with a direct and indirect death toll of nearly 100,000. About 50 cubic kilometres of material was blown into the sky. This eruption and others in Iceland may have been the cause of the "year with no summer", 1816, which led to widespread suffering and starvation in Canada, the eastern US and Europe. In that year, global temperatures were about 3° lower than usual. The 1991 eruption of Pinatubo is credited with causing a 1° temperature drop.

The most famous volcanic eruption ever was the destruction of Pompeii and Herculaneum by **Vesuvius** in Italy in 79 AD. The death toll must have been tens of thousands.

The most observed volcanic eruption ever is likely to have been that of **Mount St Helens** in Washington State in 1980, simply because the full scientific resources of the US were available to monitor it. It removed 400m from the height of the mountain, created a new 350m-deep crater and covered an area of 60 square kilometres under a landslide, probably the biggest ever observed. Pyroclastic flows and lahars caused further damage. But "only" 57 people were killed because of the advance warnings that had been issued. The ash cloud was 25km high just 15 minutes after the eruption began. For more information see:

Mount St Helens National Volcanic Monument www.fs.fed.us/hpnf/mshnvm

Mount St Helens erupting in 1980, the deadliest recent volcano in US history. Note the smaller eruptions to the left accompanying the main event.

by Vesuvius in 79 AD. We do not know what Pliny the Elder thought of it. The eruption left no trace of him but his sandals.

But volcanoes offer plenty of other hazards. For those close to the volcano at the time of an eruption, volcanic **tephra** (chunks of ash or even full-scale rocks thrown out of a live volcano) are a danger. Likewise, eruptions often cause landslides and avalanches which can catch the unwary.

But the worst volcanic hazards are those that strike far away and with little warning. For example, over 23,000 people were killed by **lahars** (mudflows) in the eruption of Nevada del Ruiz in Colombia in 1985. Lahars can run for many kilometres, often outpacing warning systems as well as people.

A volcano in action

The subduction zone where the Australian plate slides below the Asian plate has produced many of the world's deadliest volcanoes. None is more famous than **Krakatau**.

Observations of its explosion in 1883 are regarded as the foundation stone of modern volcanology. It began on 20 May, when a German naval vessel reported an 11km-high ash and dust cloud. In May and June, most of the summit of the mountain vanished. By August, loud explosions were being heard tens of kilometres away and ash was settling on ships and the nearby coastline. The Sunda strait in which the island lies was choked with pumice, the rock formed when hot magma is thrown straight into cold water. It later turned up floating in the ocean thousands of kilometres away. The culmination, on 26 and 27 August, ended with an explosion heard from India to Australia. The ash rose to over 30km high, four times the height of Everest.

The air wave from the explosion was observed all around the world. But it was the 40m tsunami it generated – far greater than that caused by the Indian Ocean earthquake of 2004 – that did most of the damage when it arrived at the coasts of Sumatra and Java. Over 36,000 people died, while the towns, villages and ships on which the survivors depended were destroyed and their farm animals were killed. Although most were caught by the tsunami, some were incinerated by a nuée ardente (see p.115), a cloud of ash and other volcanic material that crossed 40km of ocean and arrived at the island of Sumatra still hot enough to kill.

By the time it was all over, about two-thirds of the island of Krakatau had vanished. It had consisted of three volcanoes, Danan, Perbuwatan and Rakata. The first two had gone completely, as had much of Rakata. In 1927 the open sea produced by their destruction was broken when a new volcano, Anak Krakatau, the son of Krakatau, rose above the waves. It is an active and dangerous volcano. One of its explosions in 1993 killed a visiting tourist and injured five others.

Although they come from volcanoes, they do not necessarily need an eruption to get going. Sometimes they happen when a volcano breaks open the flanks of a lake and releases the water, but they can also occur when its heat melts snow and ice. Indeed, a volcano can alter drainage systems over large areas by blocking rivers as well as by releasing water.

Of all the dangers, **volcanic gases** are the most insidious. Volcanic rock contains gas even at depth, but as it rises and the pressure falls, the gas expands and helps drive the eruption. Much of it is just carbon dioxide and water vapour, but some is toxic, including sulphur dioxide, which attacks the lungs and causes acid rain. Most pernicious is hydrogen fluoride, which has sometimes fallen on cropland near volcanoes in amounts sufficient to kill grazing cattle by poisoning their feed.

But carbon dioxide is the most dangerous of these gases despite being ostensibly harmless. It kills by replacing the oxygen in the air, because it is more dense, and is especially hazardous in narrow, steep valleys which trap both the gas and its victims. Even worse, volcanic carbon dioxide can kill without needing a volcano. Lake Nyos in Cameroon has repeatedly killed people – 1700 in one episode in 1986 – by absorbing volcanic carbon dioxide filtering into it from below before releasing it in a sudden explosion. People and animals were killed nearly 30km away, as there was no volcano to warn them of the hazard. (See perso.wanadoo.fr/mhalb/nyos/.) In general, low-level volcanic activity is benign, and people have enjoyed the warm waters and fertile soils it produces for millennia, but its benevolence cannot be taken for granted.

Onwards and downwards

Volcanoes are proof that even the depths of the mantle matter to us. Now we are going even deeper, to the parts of the planet whose components – we think – never have a hope of seeing daylight. While mantle material makes its way to the Earth's surface daily, the **core** plays only a supporting role.

The discovery of the core as a specific part of the Earth is a mighty scientific achievement, depending on the painstaking analysis of earthquake records and a good deal of ingenuity. Despite its inaccessibility, we have now built up a reasonable understanding of its properties. One thing we have learnt is that just like the boundaries between the other layers of the Earth, there is no hermetic seal between the core and the overlying mantle. The core is metallic, but we know that the liquid metal that makes it up also seeps into the rocks of the mantle above.

We also know that, like the mantle, the core consists of two halves. The outer core starts about 2700km below your feet while the inner core begins some 5200km down and keeps going to the centre, about 6371km below you if you are sitting at sea level.

As we saw earlier, there is one big difference between the two. The outer core is molten. It consists mainly of iron with a few percent of nickel and about 10 percent of something else, probably oxygen or sulphur. This composition is consistent with its seismic properties and also with the chemistry of metallic meteorites. The inner core, however, is solid. Just what makes it solid despite its immense temperature is simple. It is the pressure. Even at its peak temperature of about 6000°C, the material that makes it up, with roughly the same composition as the outer core, cannot stay liquid. At the centre of the Earth, the pressure is about 14 million times the atmospheric pressure we know at sea level. The density of the inner core averages out at about 15 grams per cubic centimetre, compared to 2.7 in the crust.

It had been thought that the inner core was a relatively uniform structure. Recently, however, research has suggested that it may be only the centre of the inner core which is homogeneous, while its outer parts are more varied in composition.

Heat from the core travelling outwards through the Earth is the main source of the energy driving the movement of the mantle, although the rocks there also contain radioactive atoms of their own that decompose and generate heat.

Perhaps the most unexpected recent finding about the inner core – derived from exquisite observations of many decades of earthquakes – is that it does not share the same 24-hour rotation as the rest of the Earth, including you and me. Instead it is rotating just slightly faster, by a degree, or 20km, a year. The rotation became apparent because the whole of the inner core has a structure in which the iron crystals point north–south and earthquake waves move through them faster in this direction than east–west. That also makes it possible to spot the slight rotation of the inner core. These aligned crystals form a "fast track" for seismic waves. Over years, it is possible to observe that track moving relative to the Earth's surface by timing earthquake waves travelling from the Antarctic to Alaska.

By the standards of the solid Earth, that rate of movement is warp speed, thousands of times as fast as the pace at which continents move. And all this movement of solid and liquid metal has a very real effect: it helps to create the Earth's **magnetic field**.

This is a useful feat in several ways. As we saw in Chapter 2, it saves the Earth from the worst of the celestial radiation that would otherwise reach the surface. If it did not, life here would have taken a very different course from the one it has.

In addition, the roughly north–south alignment of the field drags any small magnetized object into line with it. The compass, first developed in China, remains a vital bit of navigational kit even in the electronic era (see box on p.123). Take a GPS receiver out into the hills if you must, but don't think of going without a map and a compass as well, and make sure you know how to use them.

The geodynamo

The machine that makes the Earth's magnetic field is called the **geodynamo**. The principles underlying it are the same as for any other dynamo. There are three components to the system: **movement**, an **electric current** and a **magnetic field**. As soon as two are present, the third will appear. In a dynamo used to generate electricity, for example, a metal wire is moved in a magnetic field, and a current is created. That is the principle behind a power station, whether the movement is created by burning coal or capturing power from the wind. In the same way, if a magnetic field is applied to a wire with electricity in it, it moves in response. This effect is used to generate force in the magnetic brakes used on theme park rides all over the world.

In the case of the Earth, there is movement, in the form of all that convecting metal, and an electric current, present inherently in the electrons in all that moving iron. And the inevitable result is a magnetic field.

The Earth's magnetic field is not very powerful. Magnetic fields are measured in units called Teslas, and the Earth's is rated at just 0.00005T. By contrast, magnetic imaging machines in hospitals have fields from 1.5T upwards, and some bizarre stars called magnetars have fields of about 100 billion Teslas. But because the Earth's core is large, the field it creates is massive as well, extending far out into space.

Our understanding of the geodynamo is being revolutionized by very large computers. Some of the biggest supercomputers in the world are being used to model the Earth's interior, such as the NEC Earth Simulator in Japan, at one time the most powerful in existence. In addition, satellites are gathering more and better data about the size, shape and strength of the Earth's magnetic field, and the core itself is being modelled – not only inside computers, but also through physical models in laboratories in which convecting masses of liquid sodium take the place of the iron found in the Earth's core.

Although the working of the geodynamo is controversial, there seems to be little doubt that it depends on the Earth having a solid inner core as well as a liquid outer one, to provide a structured shell within which metal can convect. The convecting iron seems to move at a stately speed of about 10km per year. It also seems that the north–south alignment of the convection is driven by the Earth's rotation, something we see on other planets too, including Jupiter. For more on the Earth's core and the geodynamo see www-geol.unine.ch/cours/geol/coredynamo.pdf.

Effects at the surface

Edmond Halley (of comet fame) was the first to write a scientific account of the shape of the Earth's magnetic field. He used compasses on British naval ships to map the varying direction of the field, both its north–south orientation and its "dip", the angle a compass needle at a specific point takes to the horizontal. These are the tangible signs at the Earth's surface of the way molten metal is moving far below.

The outer core convects for the same reason a boiling pan of water does: it is cooler at the top than at the bottom. The heat is produced from a number of sources, including the energy given out by iron solidifying on the outside of the inner core, which is growing slowly as the Earth cools. Add these to the force of the Earth's rotation and complex, turbulent flow ensues in three dimensions and is maintained. One result is that the field we observe at the Earth's surface is at its strongest in North America, Siberia and Antarctica, reflecting the pattern of convection cells beneath.

Switching poles

As we saw in Chapter 3, data from magnetic fields trapped in rocks shows that despite the inertia of billions of tonnes of molten iron, the Earth's magnetic field is not fixed. Instead, it can switch direction totally over a few thousand years. The last such reversal was 780,000 years ago and the average time between reversals is about 250,000 years (see diagram on p.124). But this does not mean we are "due" a reversal, any more than a gambler can be due a six just because one has not come up lately. Indeed, a look at the chart shows that at one point, there were no reversals for 35 million years.

However, we also know from direct measurement that the Earth's magnetism has become about 10 percent less strong since reliable measurements got going in the 1830s. It seems from past reversals that it takes a few thousand years for the compass needle to turn by 180°, following a period of low Earth magnetism in which it might not point very decisively in any direction.

The compass

Most of the elements in the periodic table respond only slightly to magnetic fields. But three – iron, nickel and cobalt – are **ferromagnetic**. They react strongly to magnetic fields and can retain their own field if they are exposed to one.

The reason is that the atoms of these metals have electrons which orbit their atomic nucleus in a slightly asymmetrical pattern. With most elements, the electrons are arranged symmetrically about the nucleus. A magnetic field has little effect on them because the force it exerts on the electrons gets cancelled out. With these three, there is a residual effect that can be permanent.

It seems that the Chinese were first to realize that the magnetic properties of iron might be useful as a direction indicator, probably in the eleventh century. The compass either reached Europe, or was independently discovered there, in the twelfth century. By this time both European and Chinese vessels were using it for navigation. It was the English scientist **William Gilberd** (1544–1603) who showed that for the compass to work, the Earth itself must be a giant magnet.

There are, however, plenty of factors that complicate using a compass to find your way. One is that the needle will react to any magnetic influence. On ships, this means putting an elaborate, custom-built metal structure around the compass, called a binnacle, to cancel out the vessel's own magnetism. On land, the compass will alter the direction in which it points in response to bodies of iron-rich minerals inside the Earth.

And the Earth's magnetic field can change faster than one might think, turning the compass from helper to enemy. As Alan Gurney points out in his excellent book *Compass*, as long ago as 1701 Halley published "An Advertisement Necessary to be Observed in the Navigation Up and Down the Channel of England" to alert sailors to swings in the compass that risked sending them onto the rocks.

The key to how this happens seems to lie in the detailed interaction between turbulent eddies in the outer core and the base of the Earth's mantle. This can produce patches of trapped molten iron in which the magnetism is opposite to the main strength of the field. Such a patch of "reverse polarity" might eventually spread to occupy the region of one magnetic pole while a second reversed the polarity of the other.

Richard Muller at the University of California points out that a whole range of events at the core–mantle boundary could set such changes going. These could include slumping material from old, subducted bits of continent. But he also points out that a hefty meteorite or asteroid impact would send a big shock wave into the boundary. As a result, in addition to leaving a big crater and altering the Earth's climate, such an impact

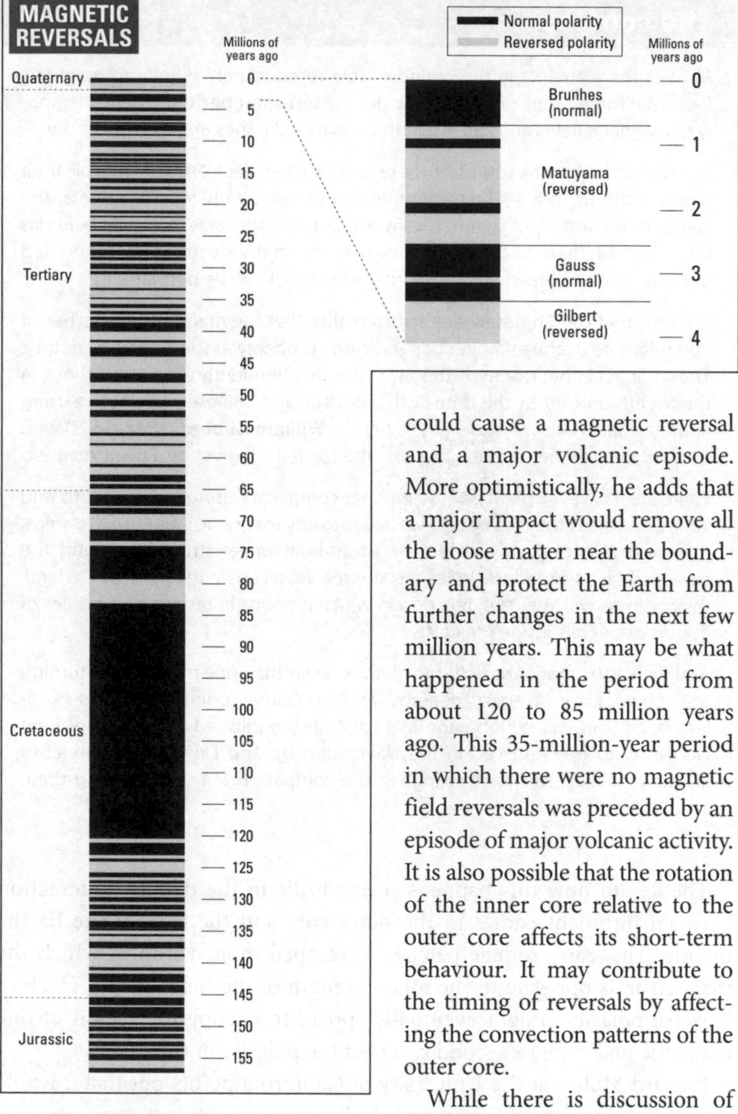

could cause a magnetic reversal and a major volcanic episode. More optimistically, he adds that a major impact would remove all the loose matter near the boundary and protect the Earth from further changes in the next few million years. This may be what happened in the period from about 120 to 85 million years ago. This 35-million-year period in which there were no magnetic field reversals was preceded by an episode of major volcanic activity. It is also possible that the rotation of the inner core relative to the outer core affects its short-term behaviour. It may contribute to the timing of reversals by affecting the convection patterns of the outer core.

While there is discussion of such a reversal happening "soon", this is soon geologically, not on any human timescale. You can expect to be using your compass on country walks for the rest of your lifetime, and to see the red end of the needle still pointing more or less to the north.

The effects of a reversal

If the Earth's magnetic field decided to reverse, just how would life change? And what could we do about it?

Start by buying shares in companies that make **GPS** receivers. For a long period before it became truly reversed, the magnetic field would be reflecting a mixed convection pattern in the Earth's core and would not be much use for navigation.

More dangerously, as we saw in Chapter 2, the Earth's magnetic field reaches deep into space and wards off some of the most harmful **solar radiation**, including ultraviolet light and charged particles. So you would not want to go out for that country walk anyway, at least not without putting on a radiation suit. If you did, the rest of the plants and animals you met would not be flourishing.

Worst off would be the birds, fish and sea mammals that depend on the Earth's magnetic field to navigate, often for **migrations** over thousands of miles. It is not likely that magnetism is the only cue they use. They also use vision, as well as clues such as the position of the Sun in the sky. But their bodies contain magnetic materials used to help point the way, and they would suffer severe evolutionary stress if it stopped working.

Differentiation

Nowadays our understanding of the solid Earth is subtle and satisfying, although the plume controversy shows that there are always new insights coming along. In particular, the idea is now firmly established that the different parts and layers of the Earth interact. Even the slow growth of the inner core has effects on the surface. Perhaps the most striking idea of the twentieth century is subduction. People have always known about volcanoes, so they had to accept that material was emerging at the Earth's surface from below. But subduction means that it is also vanishing again, and in continent-sized bites.

But even this big story misses the point. All this activity has what the literary critics would term a meta-narrative, and it is called **differentiation**. Look at the Earth today. Solid iron (mostly) at the middle. Liquid iron plus other elements on top of it. On top of that, rock. On top of that, water in the shape of oceans, and above that, air. Every time a volcano emits material to the Earth's surface, it is joining in the process of putting lighter elements on top of denser ones and so "differentiating" the Earth a little more. Despite all that crust that is recycled over time, the overall story is one of gradual differentiation.

This process has been going on throughout the history of the Earth and began when it was new, when the core and mantle separated out as the heaviest element present in abundance, iron, sank to the centre. The process is continuing as the inner core grows.

Differentiation is also going on in the Earth's outer layers. Thus the crust has more light elements such as potassium and calcium than you find in the mantle. And the sorting process continues today as lighter elements of the mantle are preferred for turning into basalt at the Earth's surface. So although the rocks you see around you in the Earth's crust have an average density of about 2.7 grams per cubic centimetre, the Earth as a whole has a density of about 5.5.

The atmosphere itself started out as the product of "**outgassing**", the gradual release of gas trapped in rocks, a process still going on today each time a volcano emits gas. Some of the gas that must have been present in the early Earth was too light to be held by the planet's gravitation and has escaped into space, so helium is rare in the atmosphere. As we have seen, the helium used by today's divers and balloonists is extracted from natural gas and is a fossil remnant of the early Earth still being differentiated out. Hydrogen, the only gas that is even lighter, tends not to escape because it is too reactive, and gets involved in forming water molecules. Water vapour, carbon dioxide, nitrogen and other gases emerge from volcanoes and would have done so in greater abundance earlier in the history of the Earth.

Differentiation is not the whole story. For example, **uranium** is dense but its chemistry means that it is concentrated in the crust and mantle, not the core, making it available for bomb-makers and reactor-builders but also supplying heat to the upper layers of the Earth as it decomposes through natural radioactivity. Another exception is **water**. The sheer amount of the stuff in the oceans is the clue. It is widely thought that at least some of it must have come from comets, abundant in the new solar system, raining down on the early Earth.

Differentiation is a continuing process. Early in the Earth's history it was at its most active, forming the structure we know today, from core to atmosphere. But on a more modest scale, it is still happening every time a volcano erupts.

5

The airy Earth

The airy Earth

You cannot feel it, but as you are reading this, a weight equivalent to a 10m column of water is pressing down on top of your head. You don't notice it because you are used to it, but it is there alright. It is the atmosphere, our subject in this chapter.

Another way to think about atmospheric pressure is to say that, from sea level to the edge of the atmosphere, there is a column of air weighing almost exactly one kilogram for every square centimetre of the Earth's surface. This means that, if you are near sea level, the atmosphere is exerting a downward pressure on you of about 1 bar. This is not the simplest measurement to take in. A bar is 100,000 pascals, which in turn is a pressure of a newton exerted over an area of a square metre.

Atmospheric pressure varies for a number of reasons. For example, the heat of the Sun makes air expand, and become less dense. As we shall see, these pressure differences drive the Earth's weather. But they are imperceptibly small to us, and the variations are measured in millibars, thousandths of a bar. To be precise, the average pressure of the atmosphere at sea level is 1013 millibar or 1.013 bar.

These variations in atmospheric pressure are measured using a **barometer**. As we have already established, air at sea level exerts about as much pressure as a 10m column of water. If the liquid were mercury, the column would be 76cm high, since mercury is much denser than water. This is the principle behind the mercury barometer. It consists of a vertical tube of mercury connected to a small reservoir of the metal which is exposed to the air. The height of the mercury column adjusts until the weight of the column balances the atmospheric force exerted on the reservoir.

Mercury is used for barometers because it is the densest liquid we have, so the barometer can be a sensible size. With a glass tube 10m or more high, you can make a water barometer, and they do exist, but you will

Living high

The higher above sea level you are, the lower the air pressure will be. On the summit of Everest, it drops to around 350 millibar, about a third of the pressure at sea level. In Rangdum in Ladakh, northern India, one of the highest places on Earth inhabited by humans, the average air pressure is about 60 percent of that at sea level.

The reduced air pressure at high altitudes means it is harder to get enough oxygen into your blood in the form of **oxyhaemoglobin**, the chemical that carries it round the body for your organs to use. No one can survive for long on Everest without an artificial oxygen supply. But much as people can adapt to living in a cold or warm climate (see p.212), they can also alter over generations to cope with altitude. Of course, the Earth's high places are also pretty chilly, and people in the Himalayas have had to adapt to both altitude and cold.

Tests on people in **Tibet**, the **Andes** and **Ethiopia** show that they make more oxyhaemoglobin than people from sea level. But these three groups, totalling about 25 million people, have adapted their physiologies in different ways to achieve the same effect. People in the Andes are often said to be noticeably red-cheeked because of their high blood flow. They also have bigger lungs than people from sea level. Llamas, which also live in the high Andes, are similarly gifted by comparison to related species that live nearer sea level.

In addition to being at a lower pressure, the air at high altitudes also has less **water vapour**. This means that people there are likely to dehydrate faster. Here such cultural adaptations as keeping your mouth closed come in handy. There is also more **solar radiation** at altitude and people there tend to have slightly less sensitive vision than those at sea level.

Andean people and their animals are both adapted to high altitude: herder and llama near Cuzco, Peru

need an impressive house to display it. The little barometers you see on walls are called aneroid, or no-liquid, barometers. They contain a very thin-walled metal box which expands or contracts as the air pressure falls or rises respectively.

Atmospheric layers

Like the solid Earth, the atmosphere is divided into distinct layers.

Take a deep breath: you have just inhaled part of the **troposphere**, the lowest layer of the atmosphere. A little like the Earth's crust below your feet, the amount of troposphere above you varies with your location. Near the Equator it is about 16km deep, but at the poles it is only half as thick.

The troposphere is a relatively thin layer, but because it is at the bottom, and has the weight of the rest of the atmosphere pressing down upon it, it contains three-quarters of the mass of the atmosphere. It is also where most of the action takes place, including the bulk of the weather. The word is derived from the Greek verb *tropos*, to turn, and refers to all the air turbulence that goes on there. (This is also the origin of the word tropic, signifying the point where the Sun turns back towards the Equator in its annual journey.)

The troposphere is defined by temperature. The average temperature at the Earth's surface is about 14.5°C. While the exact temperature you experience depends on whether you live in Egypt or Norway, what happens higher up does not. As you measure the temperature through the troposphere, it drops with height. Eventually you reach a sharp join called the **tropopause**. By this point, the temperature has fallen to about -52°C. This means a steep fall in average temperature in the tropics, perhaps 5°C per kilometre, but much less above the poles.

The reason for the drop in temperature is that the atmosphere does not absorb a huge amount of incoming solar energy, but the oceans and the land at the Earth's surface do. It is the heat that they radiate back again that is mostly responsible for warming the air. The amount of warming is at its peak near the Equator, and it is mainly air near the ground or sea that is warmed. So the temperature of the air falls steadily as you get higher.

The troposphere is only the start of the story. It contains most of the material that makes up the atmosphere, but, at a few kilometres in depth, it is thin by comparison with the layers above. Beyond the tropopause, we arrive at a layer called the **stratosphere**. Here something unexpected happens: the temperature starts to climb again, and goes on climbing until it has almost reached a balmy 0°C at the top of the stratosphere. That can mean only one thing – something is heating it up. The energy comes from the fact that this region contains the famous "**ozone layer**". The ozone layer absorbs most of the ultraviolet radiation in solar energy, and this energy warms the stratosphere. The ozone layer also protects us from the worst effects of the ultraviolet light from the Sun, although enough still gets through to contribute to skin cancer and other diseases.

Ozone, in case you are wondering, is just another form of oxygen. Chemists would call the sort you just breathed in O_2. When you breathe it in, you are not absorbing single oxygen atoms, but molecules of two oxygen atoms each. The same applies to the nitrogen that makes up most of the atmosphere. It is written as N_2. But ozone has three atoms per molecule, making its formula O_3. It is formed by a sunlight-powered reaction using two-atom oxygen molecules as its feedstock.

The Earth's coat of air, seen here from space, is a thin one, but vital to life below

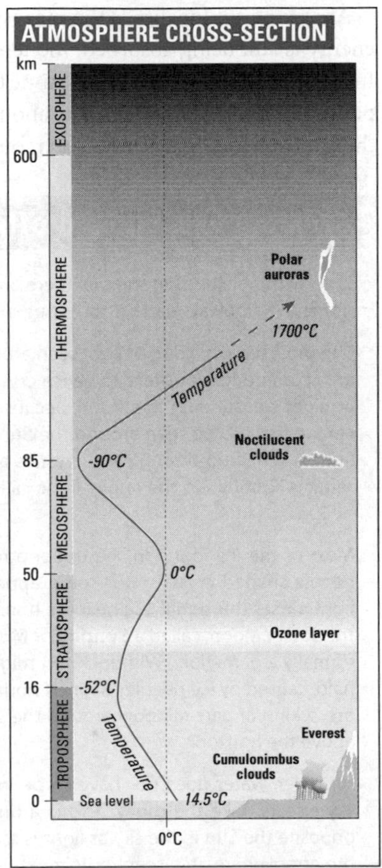

The one thing everyone knows about the ozone layer is that it has a hole in it. In fact, it has more than one, but let's start simple. The canonical ozone hole was discovered by Joe Farman, Brian Gardiner and Jonathan Shanklin of the British Antarctic Survey and exists over the Antarctic. You might think that the stratosphere is far too cold for a lot of chemical reactions to go on there. But in fact, it is so cold that ice crystals form whose surfaces are an ideal template for chemistry, especially in the summer, when there is a lot of solar energy around.

The ozone hole has arisen because of the use of **chlorofluorocarbons**, CFCs, which used to be widely used for purposes such as propelling deodorant from spray cans. Once in the atmosphere, they hang around there for many years. As they decompose under the influence of solar energy, a series of reactions generates chlorine which reacts with and depletes ozone.

The use of CFCs is on a steep downward slope, a rare and encouraging international success in environmental protection. But some other ozone depleters are still in use. In most years, the area of severe ozone depletion in spring is still larger than the continental United States. In addition, similar holes have been observed over the Arctic. On current assumptions, it will take about a century for them to heal over completely.

While the troposphere contains 75 percent of the atmosphere, the stratosphere contains a further 24 percent and extends up to around 50km above sea level. There you arrive at the **stratopause**, another join in the system.

Above the stratopause is the **mesosphere** and here, although some solar energy is still being absorbed, the temperature is on the way down again. Indeed, it will fall to -90°C by the time the top of the mesosphere – the **mesopause**, inevitably – is reached at about 85km above sea level. This high in the atmosphere there is little water, but at and around the mesopause there

Atmospheric special effects

Even with no clouds in the sky, there are plenty of amazing atmospheric phenomena to look at. Many, it turns out, are to do with ice.

The most basic is a ring of light seen around the Sun or Moon when conditions are cold enough for there to be ice crystals at altitude. It is far simpler to spot at night surrounding the Moon, because it is then against a dark sky, but with care it can also be seen around the Sun on a winter's day. It appears because light is refracted through the crystals and back to the observer. Because ice refracts light by 22°, the ring of light, called a **halo**, is always 22° from the Moon or Sun.

Most of the ice found in the upper atmosphere is in the form of hexagonal needle-shaped crystals. But some appears as flat, hexagonal crystals. When light passes through these near the horizon, a paler but still very visible copy of the Sun or Moon, called a **Sundog** or **Moondog**, will appear. (A Sundog is more formally a parhelion.) With luck you might also see a tangent arc touching the halo, caused by ice needles aligned horizontally. Rarest of all is a **circumzenith arc**, a kind of part-rainbow around the sky's zenith when the Sun is about 22° above the horizon.

However, water does not have to be in solid form to produce extraordinary sky effects. Take the **glory**. This is a ring-shaped rainbow which can appear opposite the Sun in the sky as light is scattered back from droplets of water in the atmosphere. It is seen in its most terrifying form in the **Brocken spectre** – a ghostly production in which you are bound to be the star. Any time you are walking in mountains, you may see your shadow reflected on cloud or mist below. But if a glory surrounds it, the effect is especially eerie. Certainly it used to spook people on the Brocken, one of the Harz mountains in Germany, who gave it the name.

Of course, the granddaddy of all these phenomena is the **rainbow** itself. Here the principle is one we learnt at an early age. If there are rain and sunshine at the same time, conditions are ripe for a rainbow, or more than one.

A rainbow is created when there are raindrops in the sky, and sunlight for them to reflect and refract back towards the viewer. This makes a bright disc of light in the sky opposite the Sun, but this can be hard to see. The brightest part is the bow at its edge, which shows the colours of the spectrum because light of varying wavelengths is refracted at slightly different angles by water. Blue light are sometimes seen "noctilucent" – night-shining – clouds composed of ice crystals. These are among the most difficult sights to spot in the atmosphere. They are best seen at night in polar latitudes when the sky is dark but the Sun is near enough to the horizon to light them up, in other words near dawn or dusk. (See box below for more atmospheric special effects.)

is sent back to you at an angle of 40° from the direction the sunlight arrived at, and red light 42°, so the rainbow is 2° wide in the sky, about four times the width of the full Moon. Light in this brightest bow has been refracted as it entered the water drop, reflected off the back of the drop, and refracted again as it left the drop.

Outside this primary bow is a dark area of sky, a zone from which no light can reach the observer. This is Alexander's Dark Zone, recorded by Alexander of Aphrodisias in 200 AD. Beyond it there is often a secondary rainbow caused by light reflected twice inside the raindrops. It is a mirror of the first, with the red edge at the inner side and so facing that on the outside of the brighter first bow. In ideal conditions, even-higher-order, fainter bows can be seen.

For more information and images, see:

Atmospheric Optics www.atoptics.co.uk

Sundogs and halo as seen from the Antarctic

The troposphere and tropopause are termed the lower atmosphere by meteorologists, while the stratosphere and mesosphere are the middle atmosphere. The upper atmosphere consists of just one layer, the **thermosphere**. Whoever thought of the term was a genius. The main thing in the thermosphere is heat. While its temperature rises to over 1700°C, there are almost no molecules here. Indeed, the thermosphere is generally regarded as the start of outer space. It is the region where NASA's Space Shuttle, the International Space Station and other crewed spacecraft fly, as well as a host of other satellites. The fact that there is some residual air present means that there is friction on spacecraft in orbit here. This eventually causes their orbits to decay unless they are pushed higher first.

It is almost a matter of choice where you think the thermosphere ends. At about 600km up it starts being called the **exosphere**. But this is really a rag-tag of hydrogen and helium atoms from the solar wind (see p.43) that gradually fade away into outer space after an interlude of being trapped by the Earth.

As we have seen, most of the mass of the atmosphere is in the first few kilometres above sea level. While the pressure at the surface is one bar, it has declined to 200 millibar by the time you reach the tropopause. At the mesopause it is more like 0.01 millibar.

What are we breathing?

Just what is all this air? The troposphere is extremely homogeneous. All that wind and weather ensures that it is thoroughly mixed. Unless you are standing next to a live volcano or a busy chemical works, you can rely on about 78 percent of the air around you being **nitrogen**. Nitrogen is largely inert but some plants have the means to "fix" it into living matter.

Most of the rest is **oxygen**, which makes up about 21 percent of the atmosphere. This leaves about 1 percent for all the other components. The main one is **argon**, an inert gas. It is one of the "rare" gases whose number also includes neon, krypton and helium. But when you think that it constitutes almost 1 percent of the air around you, perhaps terming it "rare" is a little harsh.

In addition, the lowest part of the atmosphere contains much smaller amounts of carbon dioxide, water vapour, various oxides of nitrogen, and other components. Despite being present only in small quantities, these substances play a vital role. They are the greenhouse gases that keep the Earth far warmer than it would be without them, by absorbing heat radiated by the Earth and preventing it from escaping into outer space. For the truth about whether we are changing the Earth's temperature by adding to the supply of greenhouse gases, see Chapter 8.

The tropopause, stratopause and mesopause are marked by severe changes in temperature gradient. But they are also the points at which the atmosphere changes its composition subtly. In the stratosphere, there is almost no water vapour and therefore very few clouds. But this layer does contain almost all the ozone in the atmosphere. The artificial chemicals that can destroy ozone are being removed from use. But other chemicals find their way into the stratosphere from the troposphere, often in huge tonnages when they are injected by big volcanic eruptions. These include nitric and sulphuric acids and their compounds.

Despite these changes in composition, the troposphere, stratosphere and mesosphere are sometimes called the homosphere because there is enough vigorous mixing between them to keep their composition very similar. But above the mesopause, there is less mixing and gravitation can slowly take effect in the vanishingly thin air there, continuing the task of differentiation with which we ended the previous chapter. Above about 200km, nitrogen dies out and the atmosphere is dominated by atomic oxygen – single atoms made by solar energy splitting up ozone and normal diatomic oxygen. Beyond 1000km we are into the exosphere, with helium dominant from 1000 to 2000km and then hydrogen.

The weather machine

Despite this reassuringly stable structure, the atmosphere is the most restless component of the Earth. Continents creep apart at a few centimetres a year, ocean currents are measured in kilometres an hour – but the fastest wind ever measured, on 12 April 1934 on Mount Washington in New Hampshire, US, was 372kph.

But although there is plenty of wild weather in the world, the climate of the Earth is surprisingly stable. Indeed, the record of ancient wind-borne sediments dating back millions of years proves that stability has been the rule for long periods of Earth history.

One major pattern of the Earth's climate that you may be aware of yourself is that most of it seems to run from west to east. This is true in the temperate regions of the Earth where, as a reader of this book, you most likely live. However, as the diagram on p.140 shows, while westerly winds (remember that a wind gets its name from the direction it comes from, not where it is going to) dominate these regions, they are not the whole story. Elsewhere, the dominant airflow can be east–west or even north–south. This is the **General Circulation of the Wind**, and is yet another thing first worked out by Edmond Halley when he was not thinking about his comet.

There are two main sources of energy for the movement of the atmosphere. One is incoming solar power. Here the story is one of change, especially on an annual basis. As we saw in Chapter 2, the subsolar point – the place on the Earth where the Sun is straight overhead – moves from the Tropic of Cancer in the northern hemisphere to the Equator, on down to the Tropic of Capricorn, and back again over twelve months. The redistribution of heat and light over this period drives the seasons.

Quicker-acting in its effect is the rotation of the Earth itself. The sight of the Sun rising in the east can also be regarded as the Earth's horizon falling below the point in the sky where the Sun happens to be. So the Earth is spinning from west to east. At the Equator, everything moves at nearly 1700kph just to manage one lap per day. But the nearer you get to the poles, the slower your rotational velocity will be, until at the poles themselves it drops to nothing. If you launched a rocket from the North Pole straight at London, it would never arrive because it would have no rotational velocity. The Earth's rotation would have shifted London out of the way before it got there. Viewed from the ground, the missile would appear to be subject to a force deflecting it to the west. We call this the **Coriolis Force**, in honour of Gustave-Gaspard Coriolis, who wrote the equations for how it works in 1835.

As we shall see, the principle had been known for centuries by that time. In fact, M. Coriolis seems to have got his name attached to a major scientific principle with less justification than almost anyone else on record.

The Coriolis Force means that Equator-bound objects, whether coming from the north or the south, are pushed towards the west, while objects heading away from the Equator are diverted east. The effect is strongest near the poles.

This is a vital consideration for anyone planning an intercontinental missile attack, or navigating an aircraft. But it also has a lot to do with how weather systems form. When one part of the atmosphere has lower pressure than another, you get wind, which is simply molecules of air moving from somewhere with high pressure to a place nearby where it is lower. It wants to go in straight lines, but the Coriolis Force means it cannot. So how do these effects turn into weather?

Spend some time in the cells

The weather is a global system and does not start in any one place, but it is simpler to pretend that it does. The best place for us to begin is at the Equator. The Sun pushes more heat onto the Earth near the Equator than it does in higher latitudes. So the air there gets hot. When air is heated, it expands and its density drops, so it starts to rise, and is pushed north or south by air following it. This rising air is wet and sheds large amounts of rain as it rises. This is why the equatorial regions are among the Earth's wettest, and are home to the great jungles of Africa and South America. Here rainfall tends to be over 1500mm a year. As my sceptical, but not very culturally aware, geography teacher Mr Ball once explained, any priest who starts a rain dance in the Amazon basin will not have to wait long before claiming success.

Because the air at the Equator is constantly being pulled upwards as the Sun heats it, it needs to be replaced. This is why the equatorial region is described as a "convergence". At the Equator, ships in the days of sail could be stuck for weeks in the "**doldrums**", a zone with little wind because the only way out for the air is straight up. But immediately to its north or south, the **Trade Winds** blow reliably – from the north-east above the Equator and the south-east below it. These winds got their name because they were supremely useful for anyone planning serious maritime commerce.

The Trade Winds are the sea-level sign of the equatorial **Hadley cells** of air circulation, identified by the English scientist George Hadley in 1735. The air that rose at the Equator is pushed north and south, and once it starts on its journey it is dragged to the east by the Coriolis Force. It also cools and inevitably starts to fall again as its density rises. This air contains too little water vapour to produce rain, making it what meteorologists term undersaturated. So the latitudes where this air descends, at the northern and southern edges of the equatorial Hadley cells, are where one finds the world's great hot deserts such as the Sahara, the Kalahari, the Mojave, the Atacama, and most of Australia and Saudi Arabia. All are about 20–30° north or south of the Equator.

However, not all the air that falls at these latitudes returns to the Equator. As it descends, it creates a high-pressure zone from which the Trade Winds head for the Equator, but the rest of the descending air heads towards the poles.

Here the jargon shifts from Hadley cells to **Ferrel cells**, and America takes over from England and France. US scientist William Ferrel pointed

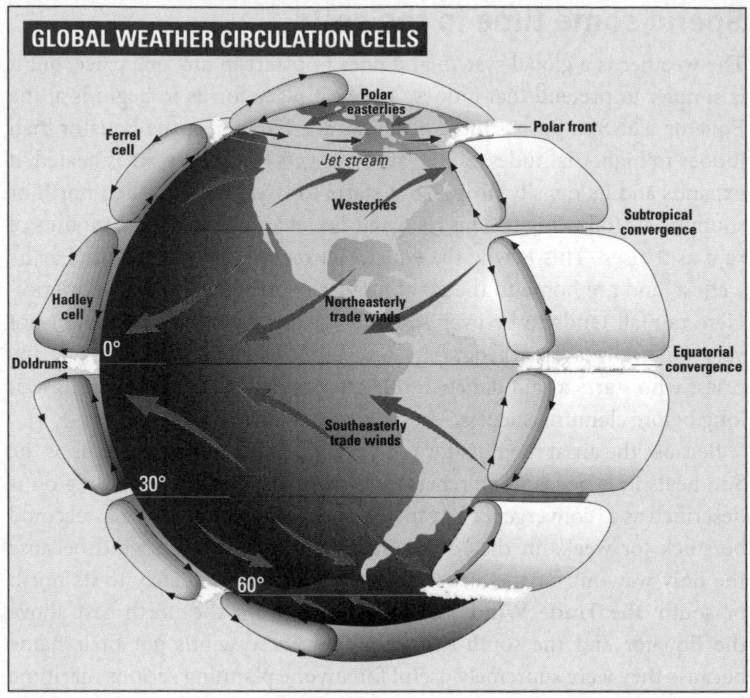

out that the temperate westerly winds that make large areas of Europe, North America and Asia so habitable by humans have essentially the same cause as the Trade Winds nearer the Equator. The combination of Coriolis Force and pressure differences means that these winds set off to the north-east in the northern hemisphere and to the south-east in the southern. As Ferrel also appreciated, and as we shall see in the next chapter, major ocean currents such as the Gulf Stream, which crosses the Atlantic from south-west to north-east, have the same origin.

The winds of the Ferrel cells are gentler than those involved in the equatorial Hadley system. As they cross the Atlantic and the Pacific, the air they carry becomes saturated with water, so that when they make landfall, the areas they encounter, such as Scotland and British Columbia, have some of the highest rainfall on the planet.

It might seem that there is nothing to stop this airflow going all the way to the North and South Poles. But it does not. The reason is that the polar regions have their own set of airflow cells, doing more or less the exact opposite of the Hadley cells.

To help remember all this, just thank the ancient Babylonians who devised the system of chopping up circles into 360 degrees. The first and hottest part of the lower atmosphere in the shape of the Hadley cells runs from the Equator to about 30° north and south. The Ferrel cells are found between about 30° and 60° in each hemisphere, while the **polar cells** run from about 60° to the pole.

The polar cells form because the extreme cold causes air to descend in the region of both poles. This air is very dry. Despite all that snow – which has built up over millions of years – the South Pole is in a desert. In the northern hemisphere, the Eismitte weather station in Greenland has 111mm of precipitation a year, less than half the 250mm generally regarded as the threshold below which an area is regarded as a desert.

This air lands and is pushed towards the Equator by more air coming along behind. Having descended at or near one of the poles, it has almost no rotational velocity and as it leaves the polar region it is heaved strongly to the west. This effect creates the easterly winds that ensured it was much easier for the Vikings to discover America than it was for them to get home again.

When these winds get to about 60° latitude, they collide with the westerlies of the Ferrel cells, forming the **polar front**. The resulting high pressure shoves the air upwards, where it divides, with half heading back to the pole and the rest towards the subtropical convergence, completing the cells.

The difference in temperature between air in the polar cells and that in the Ferrel cells produces one of the atmosphere's most spectacular phenomena, the **jet stream**, or rather many jet streams. These winds are found about 10km above sea level, blowing more or less exactly from west to east under the influence of the Coriolis Force, and at speeds of up to 300kph. Their discovery is a boon to international air travel. Planes going the same way as the jet stream go to some trouble to enter it, making the journey faster and using less fuel. By the same logic, flights heading west are routed away from the jet stream. There are also jet streams above the subtropical convergences between the Hadley and Ferrel cells. But these are far weaker, both because the Coriolis Force is less severe closer to the Equator and because the temperature differences between tropical and subtropical air are less violent than those between polar and subtropical conditions.

The explanation that this account gives of the general features of the Earth's weather machine is a good one. But it is by no means complete. In particular, the cells are not hermetically sealed off from each other.

Smaller-scale eddies and vortices cut across their boundaries, and transfer energy from the equatorial zones towards the poles. As the poles get only 40 percent as much solar energy as the Equator, this energy flow is useful for keeping the polar regions even as warm as they are.

Earth gets in the way

If things were this simple, we could save several minutes a day by abolishing the weather forecast. It would be hotter in the summer, colder in the winter, and somewhere in between in spring and autumn, in a more or less completely predictable way.

The main reason matters get complicated is the solid – and indeed liquid – Earth below all this air. The land and sea have influences on the climate and weather that we all experience. One is that air cannot flow across the land without rubbing against it. The "friction layer" or "boundary layer" extends several hundred metres above ground level. Nor is the effect simple. Depending how rugged the landscape is, there can be eddies and turbulence hundreds of metres high and wide at big landscape features, whether natural like mountains or artificial like tower blocks.

Over 70 percent of the Earth's surface is ocean, and the entire ocean surface is in contact with the air. As we have seen, air that rises high in the atmosphere and then descends to the surface contains too little water vapour to produce much rain, hence the deserts of the tropics and poles. But when air is pushed across the ocean for a few thousand kilometres, it cannot help absorbing water. In many cases it will get close to being "saturated", a state in which it cannot take in any more. However, the amount of water that a given volume of air can absorb depends on its temperature and pressure. When the wind arrives on land, things can change fast, especially if the land that it finds is mountainous. Then the air is driven uphill, where the pressure falls and the temperature does too. These factors both point in the same direction – reduced water-bearing capacity for the air, and shops down below well stocked with umbrellas. For example, Vancouver in Canada has annual rainfall of 1117mm, according to the refreshingly honest website of the Vancouver tourist authority, because the city is the target of westerly winds that have crossed the world's biggest ocean to get there. The same is true of Fort William in Scotland. Fort Bill is the small town at the foot of Ben Nevis, the biggest mountain in the British Isles, and firmly in the path of air that has just crossed the Atlantic. It has 1935mm of rain per year.

However, there is better news for anyone slightly behind the line of fire. At Inverness, on the opposite (east) coast of Scotland, and only around 100km from Fort William, annual rainfall is a comparatively bearable 640mm. This zone is called the **rain shadow**.

The many ways in which land shapes can influence weather are celebrated in names applied all over the world to local weather systems. For example, there are many variants of the type of wind known as a **foehn**, which is the opposite of the rain-dropping winds of coastal mountain regions. When that air has risen, cooled and deposited rain, it can run down the opposite slope, warming and gaining force. In the Rocky Mountains, this wind is often called a **chinook** or snow-eater as it can raise temperatures by up to 15°C in less than an hour. That sounds OK, until you see the floods that result as the snow melts.

In southern California, the wind that is steered through dry desert canyons is termed the **Santa Ana**, and it gives regions many kilometres inland a climate more akin to that of the coast. In the colder climates further north, in British Columbia, fjords facing east–west can trap a wind called the **squamish** which blows cold air in severe storms, but dies out offshore when the cliffs are not there to shape it. For more names of winds, see:

Names of Winds ggweather.com/winds.html

Another set of local weather effects arises from the different thermal properties of sea and land. The land warms up more rapidly than the water when the sun shines on it. That is why jumping in a lake feels refreshing on a hot day. In the morning, as the land warms up and the sea stays comparatively cool, the air above the land gets heated faster than the air over the sea and rises, causing a wind to blow in from the sea to replace it. In the evening, the land will cool faster than the sea, and the wind will blow towards the shore. Either way, these winds are normally mild and are called **sea breezes**.

However, the comparative properties of land and water on a larger scale are less benign. Because land masses gather heat in huge amounts in summer, and shed it in even greater amounts in winter, the centre of a continent is always a more extreme place to live, weather-wise, than the coast. From Moscow to Chicago, people who live far from the oceans get comparatively little rain, but pay for their pleasure in sweltering summers and freezing winters. Moscow is often frozen right through from November to April, but the temperature can reach 30°C in the summer.

The upshot is that for a happy weather existence, you should opt to live somewhere that is not too near the poles or the Equator, and is near to a body of water, ideally on the scale of a sea rather than an ocean. In other words, you should live near the Mediterranean.

In honour of this fact, the Mediterranean climate is regarded by the experts as being to the weather what the Mediterranean diet is to food – civilized, temperate, not too extreme, enjoyable and conducive to a long life. It is found around most of the Mediterranean, but also in and around Perth in western Australia, in northern California, in New Zealand and in coastal parts of Chile. Most of these regions are on the western edges of continents, so they get enough rain through the year for prosperous agriculture, although their summers stay dry.

Precipitation

The end result of most weather systems seems to be **precipitation** of one form or another. This is the word for falling rain, snow, hail, sleet, rime, dew and anything else that finds its way from the air to the land and consists largely of water. It appears to be untrue that the Inuit have dozens of words for different sorts of snow. But it is certainly true that people all over the world have a strong interest in the stuff the clouds throw at them and have vivid terms to describe it.

Of these myriad forms of water, **rain** is the dominant partner. Every part of the world apart from the extreme polar regions gets some from time to time and, as we have seen, some have rain more days than not.

The wettest places are simply those whose location condemns them to the most rain. The most reliable rain is that whose cause is orographic. This means that it comes about by air being pushed upwards by local topography. As we have seen, places where ocean winds meet mountainous coasts are especially prone to predictable, heavy rain. The mountain pushes the air up and as it rises, it cools down. At some point, the air will get to the point where it has more water vapour in it than it can handle. This is the "dew point". When it is reached, water vapour turns into clouds.

More common and widespread is cloud formation caused by atmospheric convection. Here the Sun heating the ground is the original mechanism. The ground heats the air above it, and the air starts to rise until, once again, it gets to the dew point.

Naming the clouds

Clouds are displays of water vapour in the sky, sculpted by flowing air. People have been observing them for centuries. The names by which we know them today were thought up by **Luke Howard**, an English scientist who lived from 1772 to 1864. Today we puzzle at terms such as cumulonimbus or cirrus, but for a gentleman of his era, Latin terminology was a *sine qua non*.

His cloud names were first set out in a lecture, "On The Modification Of Clouds", probably in 1802. As historians of science have pointed out, he really meant something close to "classification". His scheme was the rival to one thought up by a far more distinguished scholar, Jean-Baptiste Lamarck, best known for his pre-Darwinian work on evolution. One reason Lamarck's scheme failed was that the names he proposed were derived from French, but looked odd even to French eyes.

Cumulus

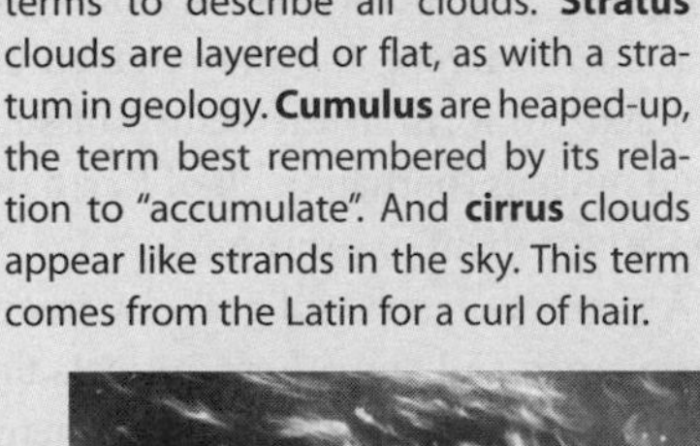

The Howard system needs only three terms to describe all clouds. **Stratus** clouds are layered or flat, as with a stratum in geology. **Cumulus** are heaped-up, the term best remembered by its relation to "accumulate". And **cirrus** clouds appear like strands in the sky. This term comes from the Latin for a curl of hair.

In addition, Howard described a type of cloud he termed "**nimbus**", from the Latin for rain. These are clouds that are producing rain or snow, or threatening to. He then added that there could be intermediate types such as cirrostratus, cumulostratus or cirrocumulus. Cumulostratus is essentially cumulus which develops to fill the whole sky.

Cirrus

Clouds more than 6000m above sea level are called "high" and are normally cirrus or close to it. Those from about 2000 to 6000m are called mid-latitude clouds and a new term, **alto**, has been added to Howard's nomenclature to describe them. They can be cumulus, stratus or a mix of the two. Below 2000m, clouds are "low".

Of these types, cirrus clouds are the least menacing. They are so high that they often contain ice as well as water vapour. Cumulus is the least agreeable as it may turn into cumulonimbus which, in addition to rain, brings the possibility of a thunderstorm. These are the deepest and highest-energy clouds, sometimes many kilometres tall and containing huge air currents. Nimbostratus clouds also produce rain but usually in a dull and lightning-free fashion. The purists regard fog as stratus that has reached ground level.

However, it is perfectly possible to have clouds with no rain. These tend to be cumulus clouds (see box on p.145), calm zones of comparatively stable air that pose little risk of rain. The rain hazard comes from larger clouds called cumulonimbus, in which convection is more active. In such a cloud, water droplets convect upwards until they become too massive for the air current below, at which point they begin to fall to the ground. Their accumulation is helped by electrical static. As the water drops rub against each other, they accumulate static charges that encourage larger drops to form.

The rule here is very simple – the bigger and deeper the cloud, the heavier the rain it will produce. If you think that a big, active skyful of clouds looks especially menacing, you are probably right, not just in the grip of the pathetic fallacy.

Snow may be rain's chillier cousin, but it is also one of the weather's genuine wonders. It forms when clouds themselves are well below freezing point. If the cloud is reasonably stable, it can contain water droplets that do not freeze and are referred to as being supercooled. Ice crystals can start to form at the centre of such a droplet, especially if there is a solid nucleus (perhaps a bit of Earth dust from below, or meteor dust from above) for them to form on.

A tiny ice crystal will grow by freezing the droplet of water in which it has formed and any others it meets until it becomes heavy enough to fall. While snowflakes are always symmetrical, they form a weird and wonderful ecology of shapes as seen under the microscope.

This mechanism explains the formation of snow, but it is also the key to a great deal of rainfall, especially outside the tropics. In temperate and polar climes, even the stuff that arrives as liquid water at ground level has often left the clouds as snow and melted on the way down. And even if it has not started out as snow, rain is usually derived from clouds that are far cooler than the ground it is falling on, and is also cooler than the air at ground level. That is why the temperature falls as rain sets in. Sometimes it is possible for supercooled water droplets to arrive as rainfall at the Earth's surface. This can form the sleek ice layers which feature in traffic hazard reports in winter.

There are some variants on these basic forms in which water arrives back on Earth. One is **hail**, which is responsible for severe damage to crops, buildings and aircraft in flight. To get it, you need clouds that retain ice droplets by repeated convection for long periods while they build up in size. The record stone (Aurora, Nebraska, in 2003) weighed about 2kg and was over 17cm across.

The process of generating precipitation is so finely balanced that, on occasion, nothing actually gets precipitated. **Virga** is rain that falls from a cloud but does not reach the ground because of the updraft below. Snow that does the same is called a **fallstreak**. Even water that falls from the sky does not have to be rain. If it is less than 0.5mm across, the term is **drizzle**, or **snow pellets** if frozen. **Sleet** is rain that has frozen in mid-flight.

It is even possible to have precipitation that does not fall from the sky. Most familiar is **fog**, water droplets formed near the ground surface as the temperature falls, typically around some sort of nucleus. The thick fogs that blanketed Sherlock Holmes' London are a case in point. Their virulence had much to do with the fact that almost every home in London at the time had a coal fire, whose smoke contained huge numbers of particles around which small water droplets could accumulate. The UK Clean Air Act of 1956 stopped the city's smog (smoke and fog) after a particularly bad one killed 4000 people in 1952.

More benign is **dew**, which forms when there is too much water vapour in the air for the prevalent temperature. When this happens – often at night – water precipitates out on some cold surface. Dew plays an important ecological role as drinking water in some parts of the world with little rain. If conditions are even chillier, it forms as ice rather than liquid water and is called **frost** or (if it is porous in structure) **rime**.

Weather fronts

When the forecast talks about rain, the word is usually accompanied by the word "front". Like the paperclip, we owe the concept of weather fronts to Norway, where they were thought up by Vilhelm Bjerknes. His early-twentieth-century Bergen School of Meteorology founded the science in its modern form.

Bjerknes's insight was that the atmosphere can be regarded as a mass of blocks of air. Within a block, the basic conditions are comparatively constant, especially the temperature and the humidity, the amount of water the air contains. But when blocks meet – at a front – the differences between them allow all sorts of mayhem to break loose.

Fronts come in several types, but the main players are cold, warm, stationary and occluded. The difference between a cold and a warm front is essentially to do with who is on the winning side, and what happens to the people below.

A **cold front** sounds the most menacing. When one appears, cold air arrives en masse in a region previously occupied by warm air. Because cold air is denser, it pushes the warm air upwards. So for us at ground level, the day suddenly feels colder. More importantly, the warm air cools as it rises. In addition, air is a fluid, not a solid, so cold air from the incoming cold front is bound to mix with the warm air it is displacing. By now you know what's coming. The warm air cools, forms big clouds, and, more likely than not, rain or even snow falls soon after. If you have not heard the weather forecast, you can still amaze your friends by recognizing a cold front coming in as you spot the first high clouds forming in a line. Because the warm and cold air have such vigorously differing temperatures, their impact can be turbulent, with thunderstorms forming – more on that below.

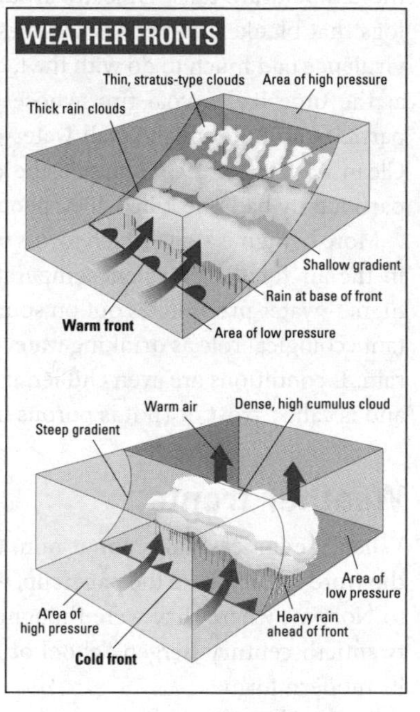

A **warm front** is slightly preferable. You are still going to need the umbrella, but you may be able to leave the overcoat at home. The warm air is less dense and is going to ride up over the cooler air whose space it is invading. As it rises, it cools, clouds form and, again, the rain follows.

In temperate climates, as we have seen, most weather runs from west to east, and the weather maps of Australia and New Zealand, or Japan, Europe and North America, typically show the fronts hurtling from left to right. However, a warm front is more likely to be delivered by a south-westerly wind – and a cold front by a north-westerly one – in the northern hemisphere, and vice versa south of the Equator. This makes sense, since air from the direction of the tropics is likely to be warm while air arriving from more polar regions is going to be colder.

Warm and cold fronts are both wind-powered and mobile. If you live in Britain, a part of the world with lively weather and brisk winds, they are a common sight on the weather chart. More unusual is a **stationary**

front, which is the result of warm and cold air meeting one another and remaining still, sometimes for days at a time. They are more common in regions with calmer wind regimes. When one sets in, clouds and rain can be generated in a single area for days at a time, so look out for rain checks at sports events and floods in vulnerable areas.

The final front we need to meet is the **occluded front**, a cousin to the stationary front. It sounds like something you might have treated at the dentist and is almost as unpleasant. Think of it as an extreme case of a cold front. In winter, you can get a warm front squeezed between cold air from both sides until it is isolated in mid-air, with cold air above and below it. The result is a region of intense turbulence and therefore of stormy weather.

These fronts are the first thing you need to look out for on weather maps and forecasts. The symbols that depict them are unvarying around the world and are shown in the diagram.

The symbology was designed in the era before colour printing was the norm, but they have survived its arrival with aplomb. A line of triangles indicates a cold front, and in colour they are blue, to suggest chilly conditions. The triangles point from cold to warm. A warm front is shown by a line of semicircles, and in colour they are shown in a warm red, this time pointing from warm air to the colder air it is likely to displace. An occluded front has them both, pointing the same way, while a stationary front has them pointing in opposite directions so you can tell which side is warm and which cold.

On weather maps in the newspaper or on TV, fronts are sometimes marked, but the basic information depicted is more likely to be limited to temperature and pressure. Wherever you live, you will soon recognize the basic shape of the weather map. In Britain, it would be a surprise not to see the hottest temperatures in the west. Even if the fronts are not marked, you can bet there is one – and work out whether it is warm or cold – by looking out for steep temperature changes, especially if they are heading in from the west or south-west. Both warm and cold fronts also need wind to push them, and therefore pressure differences, and these too are often shown on the weather map. Again, gradual pressure changes mean agreeable, balmy breezes, while steep ones mean storms.

Extreme weather

So far in our survey we've encountered nothing more unpleasant than a severe soaking, but what about when things turn really nasty? Step this way for all kinds of severe weather, starting with storms, cyclones and their even more extreme cousins the hurricanes and typhoons.

Cyclones

If you watch much news on the TV, you will be used to cyclones as the term for the tropical storms typically seen sweeping ashore and bending palm trees to the ground in a destructive manner. But if you live in temperate climes, a great deal of the weather you see is in the form of "mid-latitude" cyclones. Here, a region of low pressure around which winds are swirling inwards is a cyclone, and a high-pressure zone from which they are flowing outwards is an anticyclone.

Events here are determined by a simple balance of forces dominated by the Earth's rotation and our old friend the Coriolis Effect. It dictates that a cyclone runs anticlockwise in the northern hemisphere and clockwise in the southern.

The vital importance of cyclones to weather in the temperate latitudes was first worked out by Jacob Bjerknes, son of Vilhelm whom we met a few pages back. He grasped that the collision between the warmer westerly air of the temperate zones and the chilly easterlies of the polar regions generates much of the weather of the middle latitudes. The zone in which they meet is the polar front and the science which Bjerknes Jr developed is called **Polar Front Theory**.

The mechanism is simple and dramatic. Remember that the poles are deserts. So the air that comes from them is dry as well as cold. By contrast, the temperate regions, both north and south, have plenty of ocean for the wind to blow across and are wet as well as warm. When these air masses meet, the first effect is a slight wave in the lower atmosphere. This soon grows under the influence of the Coriolis Force, with polar air being pushed towards the Equator and to the west of the centre of low pressure, while the temperate air is pulled poleward and to the east.

These cyclones can be hundreds of kilometres across and can travel hundreds of kilometres a day, usually from west to east with the prevailing wind. The collision between these air masses is a violent one, generating storms and rain on a continental scale.

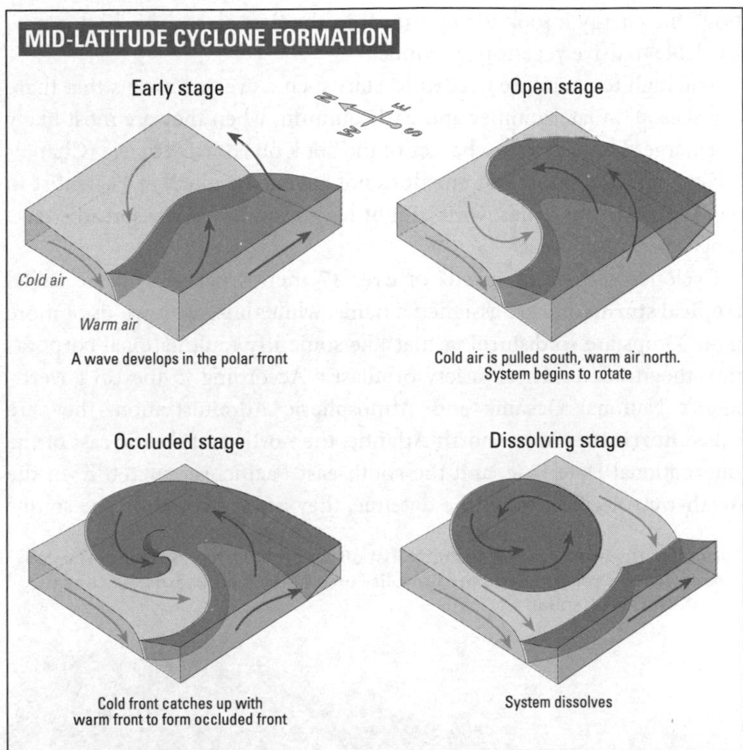

If things were this simple, the weather forecast for the temperate zones would be a task for unskilled labour. It would involve endless rainy fronts hurtling around the planet until they ran out of energy and were replaced by the one behind. The only variation would be caused by the fact that the fronts themselves are not static. They move north in the northern summer and south in the southern summer.

However, nothing in the atmosphere is quite this smooth. These temperate-region cyclones are driven by temperature differences between air masses from different parts of the globe, and they vary in size and duration. There are also cyclones in the tropics, where there are no such heat differences to drive them. Here the secret is that a sea temperature of at least 27°C is needed to get one started. Indeed, you need water at that temperature that is at least 50m deep across a large area of ocean. When this happens, the evaporation of water vapour from the sea surface builds. As the vapour rises it condenses to cloud, releasing its "latent heat of evaporation", the energy it took to vaporize it in the first place. This heat is now available to drive yet stronger winds.

The high temperature needed to start such a cyclone means that there is a season, in late summer and early autumn, when they are most likely to emerge. This is not the chapter of the book on climate change (Chapter 8 is the one you want), but one does not have to be much of a scientist to work out that a warmer world might have more cyclones, spread over a longer part of the year.

Cyclones with wind speeds of over 17 metres per second are called **tropical storms** and are assigned a name, while those with winds of more than 33mps are so disturbing that like some iffy multinational corporation they trade under a variety of aliases. According to the US government's National Oceanic and Atmospheric Administration, they are called **hurricanes** in the north Atlantic, the north-east Pacific east of the International Date Line, and the south-east Pacific east of 160°E. In the north-west Pacific west of the dateline, they are **typhoons**. In the south-

Cyclone in the northern hemisphere, just off Iceland (visible top right). The combination of low pressure and the Coriolis Force brings air towards the centre in an anti-clockwise spiral.

Contra Costa County Library
Clayton
6/28/2018 10:27:27 AM

- Patron Receipt -
- Charges -

ID: 21901024841235

Item: 31901038937209
Title: The rough guide to the Earth /
Call Number: 551 INCE

Due Date: 7/19/2018

All Contra Costa County Libraries will be closed on Wednesday, July 4th. Book drops will be open. Items may be renewed at ccclib.org or by calling 1-800-984-4636, menu option 1. El Sobrante Library remains closed for repairs.

NASA's Spitzer infrared telescope captures stars being born in the nebula RCW49 in the constellation Centaurus

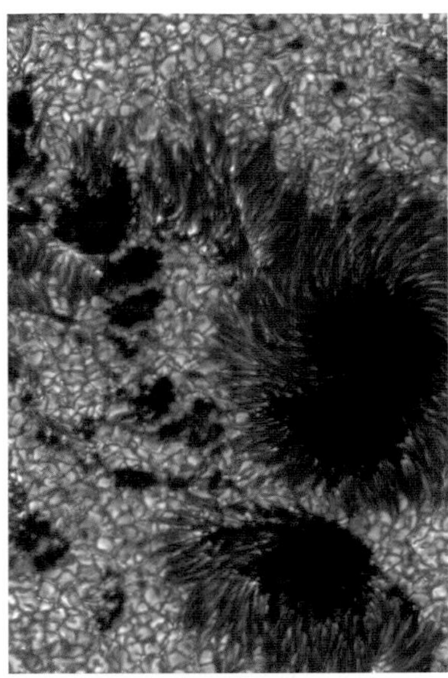

Sunspots in close-up: their magnetic fields align the charged matter that surrounds them in a distinctive pattern

The aurora over the Antarctic, seen from the space shuttle *Discovery*. The green colour is characteristic of charged oxygen atoms

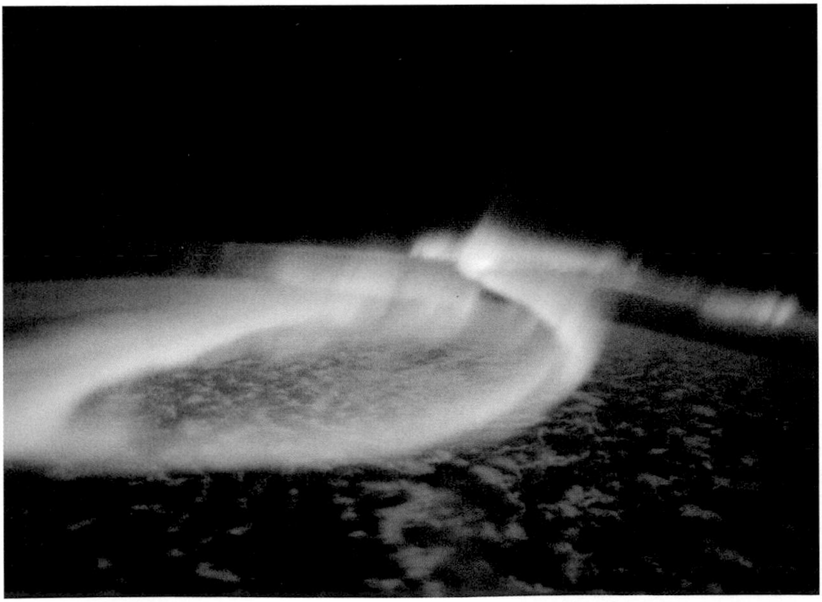

Erosion at work: The distinctive karst landscapes shown above are formed by the erosion of limestone, in China (left) and Malaysia (right). Wind and water formed the desert landscape in Utah shown below left, while the sea has eroded the coast in Port Campbell National Park, Australia, to form the Twelve Apostles (below right). These offshore rocks will be eroded away in time. The bottom picture shows incised meanders on the San Juan River, Utah

Deposition in action: The desert dunes in Namibia shown above left were deposited by the wind when its speed slowed, just as the Fraser River in Canada created the sand bars shown above right. The oxbow lake below was cut off by changes in the line of a river in the Manu National Park in Peru

Tectonics in the Earth's crust: Faulting is easy to spot in the shifted strata above left (in Oman). On a bigger scale, the San Andreas Fault in California (below left) is a hazard to several of the most important cities in the US. Folding (above right) in Cornwall, UK, is part of the mountain-building process – see the woman for scale. It eventually creates mountains such as the Himalayas (below right), seen here in a picture taken from the International Space Station, over 300km above the Earth's surface

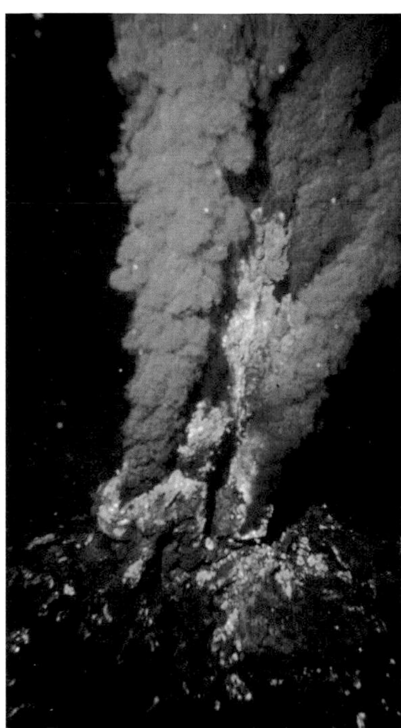

Volcanoes: Nowhere does the Earth's inner energy reach the surface more enthusiastically than in Hawaii. Above left and below, distinctive rope-like Pahoehoe lava emerges from Kilauea volcano. Above right, a black smoker is generated by heat at the Mid-Atlantic Ridge far below the ocean surface

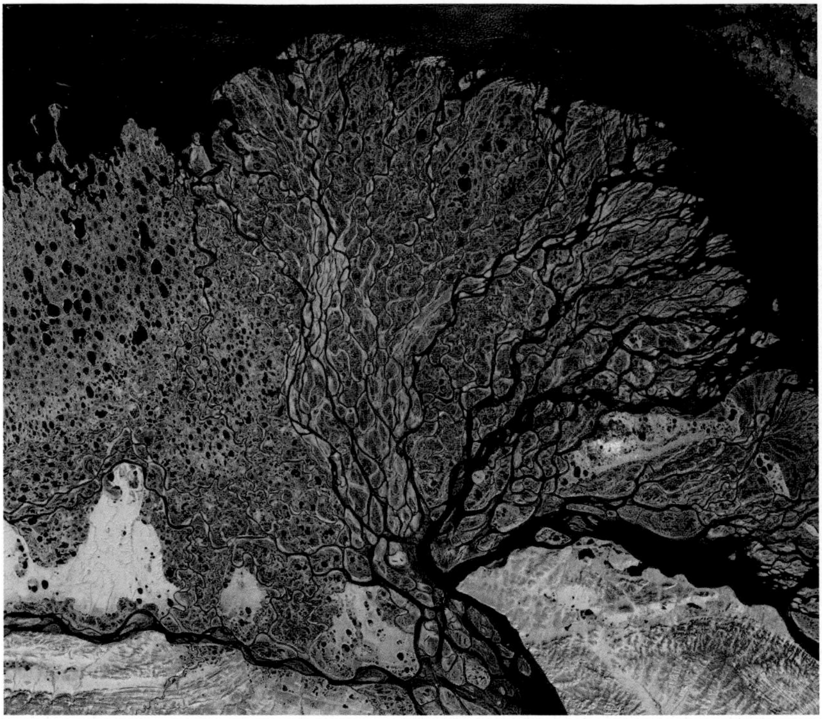

Water World: Above, false-colour image of the delta of the Lena River as it enters the Arctic Ocean in Siberia. The Lena is 4400km long and deposits immense amounts of material as it slows to enter the ocean. Below is a small part of Australia's Great Barrier Reef, the biggest living structure on Earth, made by coral but home to many other species including hundreds of types of fish

The icy Earth: Above left, tundra in the Richardson Mountains of Northern Canada. Above right, the Ross Ice Shelf of the Antarctic, which ends in a cliff up to 50m high. But remember that almost all of the ice is below sea level as the shelf is afloat. Below, the patterns of rocky moraine on the Meade Glacier in Alaska are formed by rocks falling from mountains and being swept to the centre as glaciers merge

Human impacts: Most people live in cities – this one is Toronto in Canada – but urban living does not have to deplete the planet. Ways of getting food from the land include traditional, labour-intensive methods such as these contour-hugging rice paddies (below) and low-labour, resource-hungry ones such as these circular fields in Kansas, US (bottom). The circles are signs that rotating irrigation equipment is in use, and the colours show different stages of crop growth. The Landsat satellite that took this image is descended from US spacecraft originally designed to spy on Soviet agriculture

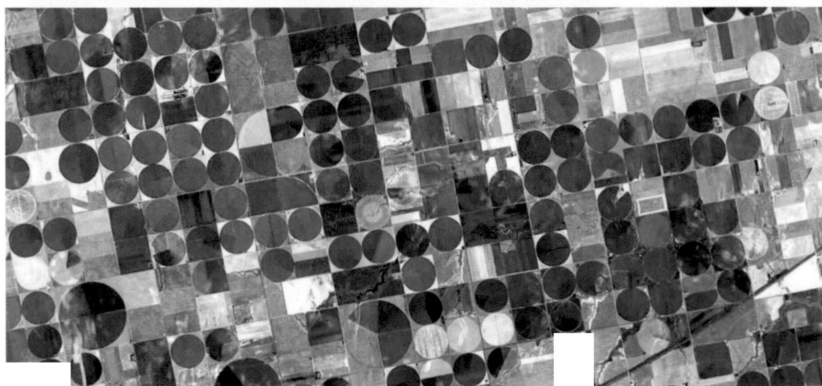

Naming the guilty parties

As if the plethora of terms used to describe tropical storms weren't enough, the most significant storms are assigned their very own name too. But where do these names come from? Namer-in-chief is the World Meteorological Organization, which controls the lists of names. But local bodies such as the US National Oceanic and Atmospheric Administration actually hand them out. There are six lists of names for hurricanes in the Atlantic. Before 1979 they were all female, but men were added in response to increasing feminist complaint. The lists come round every six years, so the one from 2007 will be used again in 2013. But names are added and removed, and a name used for a highly damaging hurricane, especially one with a large loss of life, can be retired to avoid reviving bad memories. Examples include Andrew, a famously damaging hurricane from 1992. The names are kept simple, with the likes of Tommy and Wendy much in evidence. If you are called Gustavus-Adolphus, there will never be a hurricane named after you.

Similar systems are used to name storms in other parts of the world, and the practice clearly helps raise public awareness of the approaching danger, as well as making communications between professionals simpler too.

Storm Names www.wmo.ch/web/www/TCP/storm-naming.html

west Pacific west of 160°E, and of course the south-east Indian Ocean east of 90°E, they are **severe tropical cyclones**. In the north Indian Ocean they are **severe cyclonic storms**. And if you see one trading as a **tropical cyclone**, you are in the south-west Indian Ocean. Quite how this variety has lasted in the era of the Internet is a mystery, but harmonizing it would probably take more effort than bringing peace to the Middle East.

The monsoon

Nowhere are annual variations in weather more extreme than in the Indian Ocean, home of the **monsoon**. In the northern hemisphere summer, the monsoon wind blows from the south-west, and is a wet wind whose rain is vital to life across India and adjoining countries and which deposits metres of snow across the Himalayas. The severe weather that the monsoon brings is the ultimate orographic rain. If the Himalayas were not there, the weather would not be trapped and the wind would blow into central Asia instead of depositing rain and snow en masse in and around India. As we have seen, the weather systems shift south with the Sun during the southern summer, and this wind is replaced by one from the north-east, which is dry as it has passed over the land. It brings with it the lengthy dry season which, along with the monsoon rains, characterizes the climate of India and the neighbouring region.

The monsoon is a form of extreme weather whose predictability makes it a valued part of the Earth's normal variation. As we shall see in Chapter 8, there are suggestions that the whole climate system that drives it is being altered by human behaviour. Certainly the possibility of the monsoon failing amounts to a mortal threat to life in western Asia. For more information see:

Monsoon Online www.tropmet.res.in/~kolli/MOL

Deadly weather

Despite the ordered pattern which the weather tends to follow, its extremes do not exist merely to provide footage for those late-night "when storms attack" programmes on the higher-numbered TV channels. Indeed, bad weather is the biggest killer after earthquakes among natural disasters ranging from volcanoes to meteorite strike. (People have been injured by meteorites but there are no reliable records of even a single fatality; see www.ilankelman.org/disasterdeaths.html for more on deaths from natural disasters.)

A single event, the 2005 flooding of New Orleans by **Hurricane Katrina**, makes the point. The economic damage was priced at about $200 billion, but the floods killed "only" 972 people in the official count. The decision to attempt to protect a major world city from flooding when it was sited below sea level and located between a river, a lake and a sea noted for extreme storms was bound to come unstuck at some point. If a developing-world country had done the same thing, the financial cost would have been less because Americans have the most money. But the cost in lives would have been far greater. Indeed, anyone surprised at the fate of New Orleans has not looked at the lessons of history along the US Gulf Coast. In September 1900, a similar storm flooded Galveston in Texas, some 500km from New Orleans. At that time, both sea defences and warning systems were less developed. About 6000 people were killed.

Although too much water is the cause of many immediate disasters, having too little has killed more people over longer periods of time. In the US, the **drought** of 1988–89 was blamed for $40 billion of economic damage even though it killed few if any people directly.

In the developing world, NASA points to the drought of 1991–92 as one of the most serious, affecting 24 million people across 6.7 million square kilometres of Africa. In 1984–85, famine resulting from drought across the Sahel, the part of Africa at the southern boundary of the Sahara, killed about 800,000 people, perhaps three times the death toll of the worst earthquake on record.

There are many definitions of a drought. Some are very practical, relating to having too little water for crops. Others are more theoretical, and refer to how long rainfall has dropped by some percentage below normal. You can't have a drought in a place that is already a desert, because it is defined as a spell when the weather is drier than normal.

But droughts tend to have a simple set of causes. They are normal weather taken to extremes. One pattern of drought weather, often likened to the formation of an eddy in a flowing river, occurs when an air mass gets separated out from the general circulation of the atmosphere and sits in a particular spot for a long time. It may start out by dumping large amounts of rain. But once it is out of water, no more can fall and the drought will continue until such a "blocking system" drifts off or gradually rejoins the rest of the planet's airflow.

For more information on droughts, see:

National Drought Mitigation Center www.drought.unl.edu

Lightning

In the US, according to the National Oceanic and Atmospheric Administration, lightning kills an average of 66 people a year: more than volcanoes or hurricanes, fewer than floods, but still 66 too many. But lightning is one of the most preventable deaths. The solution is to either get indoors (and stay away from wires and windows) during a thunderstorm or, if you cannot manage that, lie down. The electricity is looking for the fastest route between Earth and sky and will ignore you if it can find a better way.

Lightning forms in turbulent clouds containing a lot of ice. The particles rub against each other and collect opposite charges of static electricity. The positively charged ice accumulates nearer the top of the cloud and the negative near the bottom.

Many lightning strikes occur within or between clouds, connecting and equalizing these areas of opposing charge. Those that hit the ground do so because the negative charge in the cloud attracts positive charge on the ground below until a conducting channel opens between the two. Then two things happen. A big current – lightning – flows through the temporary conducting channel of air. And the heat of the passing current expands the air so fast that it generates a shock wave that you hear as thunder.

As we will see in Chapter 8, there are suspicions that both floods and droughts are increasing in frequency as the Earth gets warmer. But much weather damage is caused by smaller-scale effects. For example, the **tornadoes** of which about 1000 a year sweep across the US are a highly specific weather phenomenon. They are defined as hanging from a cumuliform cloud at the top, and touching the ground at the bottom. They arise from rotating thunderstorms in areas where there are major temperature differences between adjacent bodies of air, but even the US government scientists who have devoted huge resources to finding out about them admit that they hold many secrets and often fail to appear when conditions seem promising.

Record breakers

Even places with dull weather can break records. In 1893 Mile End, then as now a bustling part of east London, and used to its fair share of wet, set the UK record by having no measurable rain for 73 days from 4 March to 16 May.

However, the real records are set nearer to the poles and the Equator. Arica in Chile has the least rain in the world, at about 0.76mm a year. This is under one ten-thousandth of the amount falling at Kauai in Hawaii, 1168cm a year.

Arica is in the Atacama Desert which, as we saw, is caused by descending hot, dry air that started its journey at the Equator. It turns out that most weather records, whether for heat or cold, or for dry or wet conditions, arise not from freak events but from normal weather that gets slightly more out of hand than usual. For example, extreme cold kills over 700 people a year in the US. It normally kills people who live in areas where there is extremely cold weather every winter. Extreme heat kills an average of 384 people a year in the US, again mainly in places where it is a predictable, and predicted, part of the annual weather pattern. Despite the attention paid to storms and hurricanes, it is cold and heat that cause the greatest loss of life. In the US, storms kill about 50 people a year, floods about 80 and hurricanes and tornadoes another 70.

The coldest place recorded on Earth is the Russian Vostok base in the Antarctic. It went down to -89.2°C on 21 July 1983, a month after midwinter's day. The heat record is held by Al Aziziyah in Libya, where it hit 58°C on 13 September 1922, a degree ahead of Death Valley in California on 10 July 1913.

The least gripping weather

Most big news stories about the weather are to do with extremes. Snow falling metres deep, hailstones the size of footballs, winds faster than jet planes: these are the stuff of weather legend. But spare a thought for one record that is too rarely explored. What about the record for crushingly dull weather?

Spare a thought for the Englishwoman who moved to **Arizona** a few years ago. On the first Monday in her new job, she exclaimed to her colleagues: "What a nice sunny day." She did the same thing on Tuesday and on Wednesday. Then someone finally told her that everyone thought she was mad. It is sunny every day in Tucson, they told her, and on the rare occasions when it is not, you can say something about it.

Well, Tucson weather is indeed pretty dull, with the temperature wafting from the teens to the thirties and back again in a highly regular fashion. There is almost always about 250mm of rain, mostly in the autumn. But in recent years, the city's claim to true dullness has been blighted by an almost newsworthy drought.

Another candidate might be the **South Pole**. As a desert, it has only a few centimetres of rain – or its snow equivalent – per year. The blizzards there mostly pick snow up, throw it about, and then drop it again. Although the temperature varies from, say, -60°C in winter to -30°C in summer, both are well below freezing anyway.

However, to find true climate dullness, it is necessary to head into more equa torial regions. In the **Amazon**, it is completely typical for clouds to build through the day until a hefty storm gets you in the afternoon. Although some times of year are rainier than others, it rains all year round. The Amazon basin is also pretty homogeneous in terms of weather despite being bigger than most countries.

But the climatic dullness record is held by **Jakarta**, capital of Indonesia and of unremarkable weather. A look at the data shows that you should expect "discomfort from heat and humidity" in January, December and all intervening months. In April and May, the average minimum temperature is 24°C, while in the other ten months, it slumps to 23°C. The average maximum ranges from 29°C to 31°C. Even the records since data gathering began only range across 2°C for the lows and 3°C for the highs through all twelve months of the year.

The reason for this stability is that the equatorial regions are marked by a strong upward airflow. Because the area is heated at more or less the same rate by the Sun all year round, conditions are very constant. There is a rainy season and a dry season, but even these terms are comparative. The rising air cools as it ascends and a decline in its ability to hold water is inevitable. The result is rain for those below. There are plenty of places where the weather is very predictable, such as the centres of the American or Eurasian land masses, where hot summers and freezing winters are almost the rule. But the tropics are the place where variation in the weather is all but unknown.

The most varied weather

We may have identified the places with the highest and lowest temperatures, and the strongest winds. But where in the world sees the most variety in its weather forecasts?

Finding the biggest combination of weather extremes is tricky, but there is no doubt that the bigger your country, the more weather it can have. Smaller countries such as the UK and Japan, parked on either end of the Eurasian landmass, have big seasonal variations. But the US and Russia both have glaciers as well as deserts, and deep snow as well as baking heat. Your author's attempts to find Russia's most variable weather spot have come to naught. So salute Springfield, Missouri, holder of the weather variation record for the US. Combining measures such as temperature, wind, rain, thunder, humidity and fog, it comes out ahead of everywhere else in the US for weather variety. Part of the reason is that it has no mountains to protect it from freezing air from Canada, or from warm air heading up from the Gulf of Mexico, and it is also in the tornado zone.

For more on this and much else weatherish, see:

Weather Pages www.weatherpages.com

Planes, boats, skis

Despite the attention paid to the exceptions, much weather is a story of modest variation around the average. There are, however, times and places when the details are all-important.

Possibly the first profession to find this out were **sailors**, for whom the weather has long been a prime hazard. For most of history, too, the weather has been the engine powering most of the world's shipping.

The weather at sea is simplified by the absence of land, so the effects of air striking mountains or other land masses are removed. But the added complication is that the land usually stays still when the wind blows, while the sea reacts to the wind blowing across it.

The result is that sailors need ways of describing how much trouble they are in, and that cover both water and air. As long ago as 1805, the British admiral **Sir Francis Beaufort** developed a way of quantifying the wind that is now in use worldwide. He defined criteria for recognizing wind speeds both on land and at sea. His scale starts at 0, "calm", characterized by smoke rising vertically from chimneys or, at sea, "sea like a mirror", but soon reaches categories such as 7, "near gale", with whole trees in motion on land and white foam being blown in streaks at sea. It concludes with

11, "violent storm", which is "accompanied by widespread damage" and, at sea, small and medium ships vanishing from view behind the waves, and finally 12, "hurricane", which is simply worse than all the other grades, with wind speeds above 72mph. Note that in each case, more severe wind means more severe waves.

In addition, sailors need to know what they are looking at, since if their ability to see declines, so does their safety. "Fog" is defined as seeing a mile or less, "poor" visibility 1–3 miles, "fair" 3–6 miles and "good" anything better.

When **flying** came along, a new industry whose safety and ability to operate depended on the weather was created. In Britain, the Meteorological Office was for many decades more or less a front for the Royal Air Force, funded from the defence budget, much as the Army had the Ordnance Survey, which mapped the land.

Pilots have to worry about the general circulation of the atmosphere, to avoid flying into headwinds too often, but must also know about more immediate problems. One is seeing where they are going. Although radar and satellite navigation have ended the days when fog could ground all planes, pilots still like to know the "ceiling", or the height of the base of continuous cloud cover, that they are flying into. The key distinction is between instrument meteorological conditions and visual meteorological conditions. If there is enough visibility, a pilot can land a plane by brain power, but if there is not, instruments are needed as well.

The most specific flying hazards posed by the weather arise from thunderstorms. Radar can spot them at a good distance, and it is possible to steer round modest ones. The most dangerous are large storms which can generate severe updrafts and downdrafts of over 1km a minute. Light aircraft are warned to avoid them by up to 20km because of their ability to spawn new storm cells. Other prime aviation hazards include "microbursts" of cold air hurtling straight down from a convecting cloud – just the thing as you are taking off or landing.

The other group who need some specialist knowledge of the weather includes anyone who spends serious time a long way above sea level without using an aeroplane. **Mountains** make their own weather, and it is usually less friendly than conditions down in the lowlands.

The first problem with mountains is that as you climb them, the temperature drops. A warm sunny day in the valley may be a cold, bright day 1000m up. That is fine if you are ready for it, but not if you have set off in the clothes that made sense at sea level. Many similar but slightly differing estimates are published for the average temperature drop with height. They average out at around 6°C per 1000m climbed, but that is an average

Shepherd's delight

The adage "Red sky at night, shepherd's delight; red sky in the morning, shepherd's warning" has high-powered support. Jesus mentions it in the Bible (Matthew 16, verses 2 and 3), and it appears in Shakespeare's epic poem *Venus And Adonis*. But how does it work?

Remember that the sky is blue. This is because the air scatters bluer light from the Sun more than it scatters light of longer, redder wavelengths. So if you see a red sky at sunset, it means that sunlight has travelled a long way in the atmosphere. This shows that the air is very clear to the west, where the Sun sets. Because most weather patterns move from west to east, that means that calm weather is probably on its way to you.

But if you have a red sky at dawn, the clear air must be to the east. This means that the calm weather has passed. It is a good bet that conditions for shepherds will soon worsen as bad weather arrives from the west to replace it.

and, especially in winter, temperatures can drop much faster than this. As well as an increased risk that you are going to get cold, this means that ice and snow can persist on mountains, especially in spots that get no direct sunshine, long after they have vanished elsewhere.

Next, the air itself has to climb as it hits the hills. This means snow and rain, but it also means cloud, as the air rises and cools and the water vapour it contains condenses. The classic case is Table Mountain in Cape Town, South Africa, where the "tablecloth" of cloud is a frequent sight on top of the hill. It looks lovely, but such cloud can be less welcome if you are having trouble finding your way down.

Table Mountain and tablecloth, Cape Town, South Africa

Nor are you completely safe away from the mountain tops. The air that does not find its way over the hills can be forced through a narrow valley instead, making for unexpectedly high winds.

On the up side, the distinctive weather of mountain areas is now the subject of a wealth of specialist forecasts designed to lessen the annual cull of walkers, climbers and skiers. Almost every mountain region has one and there is no excuse for not consulting it before you set out for the hills.

Whatever next?

The weather is a severe hazard in some parts of the world, and even where it is depressingly familiar, people love to complain about it. But in recent decades, their discussion has at least become a lot better informed, because our ability to forecast the weather has been much enhanced. Perhaps more importantly, forecasts have also become more useful to everyone from farmers to planners of major sporting events. According to the UK Meteorological Office, the three-day forecast today is as good as the one-day forecast was twenty years ago.

The simplest method of forecasting the weather is to produce a "**persistence forecast**". Pay attention and I will teach you how to do this, at no extra cost. If today is rainy and the temperature at noon is 17°C, the persistence forecast for tomorrow is rainy with a noon temperature of 17°C.

This sounds idiotic because it would allow you to make a forecast with no knowledge of meteorology at all. But it is not. To make a valid persistence forecast, you have to know that you are in a part of the world and a part of the year where the weather is very stable.

Forecasts have improved beyond the surprisingly high level of accuracy of persistence forecasts because of a fortunate conjunction of two factors. The first is that we have far **more data** than before on the current state of the atmosphere and on its historical condition. It is gathered from the Earth's surface, both on land and at sea, as well as from balloons and aircraft, and from space.

This data is so important and so perishable that its collection and free availability on a world scale are overseen by a UN agency, the World Meteorological Organization. As its website reports, the total amount of infrastructure in play includes "some 10,000 land stations, 1000 upper-air stations, 7000 ships, about 1200 drifting and moored buoys and fixed marine platforms, 3000 commercial aircraft, six operational polar-orbiting satellites, eight operational geostationary satellites, and several environmental R&D satellites/space-based sensors".

Although weather data is now being gathered on a far larger scale than in the past, just what is gathered has changed little. Measurements of temperature and pressure are probably the most basic, followed by the humidity of the air, wind direction and, on the ground, the amount of precipitation.

The other part of the picture is that **increased computer power** has allowed this data to be used to create genuine forecasts. In many countries, the state meteorological body has the largest computer in the land. The largest Linux database in the world, in Hamburg, Germany, is devoted to climate data, while the Earth Simulator in Japan was for some years the world's fastest computer (see also p.121 for its use in computing properties of the deep Earth).

The basic way in which a weather forecast is made is simplicity itself. You get as much data as you can on the present state of the atmosphere and use basic concepts, such as the flow of air from high pressure to low, to see how different it will be some time later. To do this you chop the atmosphere up into blocks, in three dimensions, and specify the initial condition of each. Then you let air in the boxes interact according to the laws of physics and see what the state of play is in, say, an hour or twelve hours. Then you can repeat the process, using the result of the first step as input to the next one, to generate an even more distant forecast.

This approach was thought up long before computers existed, or the data to feed them, and weather forecasts have long been one of the drivers of big-league computing. For a computer system to generate a useful forecast, it needs to run faster than the weather itself. A perfect forecast of the weather in 24 hours is of limited value if it takes 48 hours to appear.

Computer models have revolutionized weather forecasting. But if you set a computer running on a set of data describing the atmosphere as we see it now, it would still be foolish to think it will produce a perfect forecast for all time. In practice, even today's forecasts drop fast in reliability when they get more than a week ahead. This is described as the forecast "losing its skill". Part of the difficulty is that the input information for the forecast is imperfect, but a more basic objection is the "**butterfly problem**". This is shorthand for saying that the weather is a chaotic system. It is not chaotic in the way that the top of my desk is chaotic. In this context, "chaotic" means that it is never possible to allow for every minor immeasurable fluctuation in the input conditions, so that a butterfly flapping its wings over the Amazon could in the end produce a storm in New York.

Even in the shorter term, weather forecasts fall down because the computers that produce them can generate only a central expectation if they are fed only a single set of starting data. One way round this is to set a number of forecasts running with slightly different input assumptions. This is an "**ensemble forecast**". If five are run and they all agree that tomorrow will be dry, that gives the forecast more credibility. If they predict everything from blizzards to Saharan sunshine in a few days, it is time for a rethink. In practice, some ensemble forecasts retain at least part of their skill for up to fifteen days ahead. In the Canadian weather forecasting system, sixteen forecasts with varying assumptions are run in tandem, while the European Centre for Medium-Range Weather Forecasting manages 51, a central prediction plus 50 variants.

As well as adding to the credibility of the central forecast, these ensemble forecasts allow us to get an idea of how likely variation is around it. So a single forecast will predict either rain or no rain. Run enough alternative versions, and you can produce those handy predictions that there is, say, a 60 percent chance of rain.

6

The liquid Earth

The liquid Earth

No popular description of the Earth can get far without using the term "blue planet". Pedants might object that the solar system contains another planet – Neptune – that is also blue and is four times the size of this one. But the Earth has no plans to relinquish the title.

The Earth's distinctive blue colour is to do with water. Throughout its history, most of the Earth's surface has been under water. Look at it from space and the main thing you see is water, which reflects blue light while absorbing the redder light at the other end of the visible spectrum. Pictures of the Earth from orbit or from further afield also show land, which looks mainly brown, and clouds and ice, which are white. But it is water that dominates the scene because it adds up to over 70 percent of the total surface area of the Earth.

In this chapter, we shall look at water in several different ways. It is essential to life. It is a natural resource that people are using in increasing amounts. And it is arguably the most significant player in shaping the Earth's surface and driving the processes that make the Earth the place it is.

Water, water, everywhere...

Water exists almost everywhere on the Earth. In the oceans it can be thousands of metres deep. But it also occurs in the form of ponds and puddles (which we shall not dwell on too much), seas and lakes, streams and rivers, and ice caps and glaciers, the subject of our next chapter. Water makes up a significant percentage of the weight of all living creatures. Only plants and animals that have become specialized in living in dry conditions by long millennia of evolution can last long without a drink.

Why is water so important? The answer lies in its physical properties, which would be astounding if they were not so familiar. Across the range of temperatures found at the Earth's surface, it can exist as a vapour (water vapour), a liquid (water) or a solid (snow or ice). The changes in phase between water and water vapour, driven by solar energy, are the key driver of the weather, as we saw in the previous chapter. In addition, a water molecule has positive charge at one end and negative at the other (it is covalent, in chemistry-speak), which makes it a brilliant corrosive and solvent, constantly eroding away any material it meets and transporting it somewhere else. Not for nothing is there a spoof description of the Earth, written by a supposed visiting alien, claiming that the place is totally uninhabitable. Most of its surface is covered by this ferocious solvent while its atmosphere contains fatal amounts of the equally reactive and unpleasant chemical oxygen.

So how much of this desirable substance does the Earth actually contain? One recent estimate puts it at 1385 million cubic kilometres. If the Earth were perfectly smooth, this would provide enough for a sea 2700m deep covering the whole planet. This would make a fair swimming pool, but is trivial compared to the amount of rock that lies beneath, as the Earth's radius is just over 6300km.

...Nor any drop to drink

Of this water, the vast majority is in the oceans and about 97 percent of it is saline. That means that it contains about 3.5 percent dissolved material, most of it common **salt**, sodium chloride. The usual technical word for something salty is saline. But because the salts dissolved in sea water include others beside sodium chloride, scientists prefer the term haline, for salts in general. In addition, a look at the detailed composition of sea water confirms water's powers as a solvent. Pretty well every element in the periodic table is there, albeit sometimes at a level of only a few parts per million. The solvents used in processes such as dry cleaning cannot even begin to compete. They are used only because they are better than water at leaving fabrics looking nice after cleaning, and at dissolving some organic chemicals.

Because most of the Earth's water is salty, the 3 percent that is not is especially valuable, not least to human beings, for drinking and for other activities that need pure water. But much of this non-saline or "fresh" water is locked up in ice caps and glaciers, so the amount of available fresh water is even less.

The saltiness of sea water is far from constant. The salt and other dissolved material that it contains has mainly got there by river after being eroded from the land, although there are other sources such as undersea volcanoes. Sea areas like the Baltic, which is connected to the rest of the world's oceans by only a narrow channel, are less salty than the ocean at large. The Baltic gets most of its salt from occasional bursts of ocean water that make their way from the North Sea in storm-driven "saline pulses".

At the other extreme, areas of the ocean that lose water are likely to become steadily more salty. For example, if you freeze salt water, the solid ice that forms is not salty, at least until you freeze the last little bit and the salt has no choice but to follow. This is good news for polar explorers, who can melt ice for drinking water. But it also makes polar waters saltier in the winter. By the same logic, sea water in the tropics is saltier because more water evaporates from the sea at higher temperatures. So salinity is at its lowest in temperate oceans and rises as you travel either north or south. In the same way, river water is less saline than ocean water, so sea water near the mouth of a major river will be less salty than water from mid-ocean. While the average salinity of sea water is about 3.5 percent, it falls to below 3 percent near major rivers. Water in the 1–3 percent salinity range is termed **brackish**. These lower ranges of salinity tend to occur in tidal estuaries. If the salinity goes over 4 percent the water is termed **hypersaline**. The best-known case is the Dead Sea, which is filled by rivers but empties only by evaporation, so the salt that finds its way there has no way out.

The hydrological cycle

The big message of this book is that the Earth is a single, understandable system in which everything joins together. Nowhere is this more apparent than when we consider the **hydrological cycle**. This is the machine that connects the Earth's airy, watery and icy zones. In these pages, these parts of the planet are separated out because mingling them in a single chapter would have led to a muddle. But a water molecule that is in the Atlantic today could tomorrow form part of a snowdrift in Greenland, or be carried along in a stream in France.

The hydrological cycle is not a completely closed system. Some water molecules do escape the Earth's gravity into space, while new ones arrive with cosmic impacts (a big comet like Halley contains hundreds of billions of tonnes of water). But these processes occur at an immeasurably slow rate. So the cycle is more or less self-contained. This means we could start describing it at any point. Let's stick with tradition and start at the

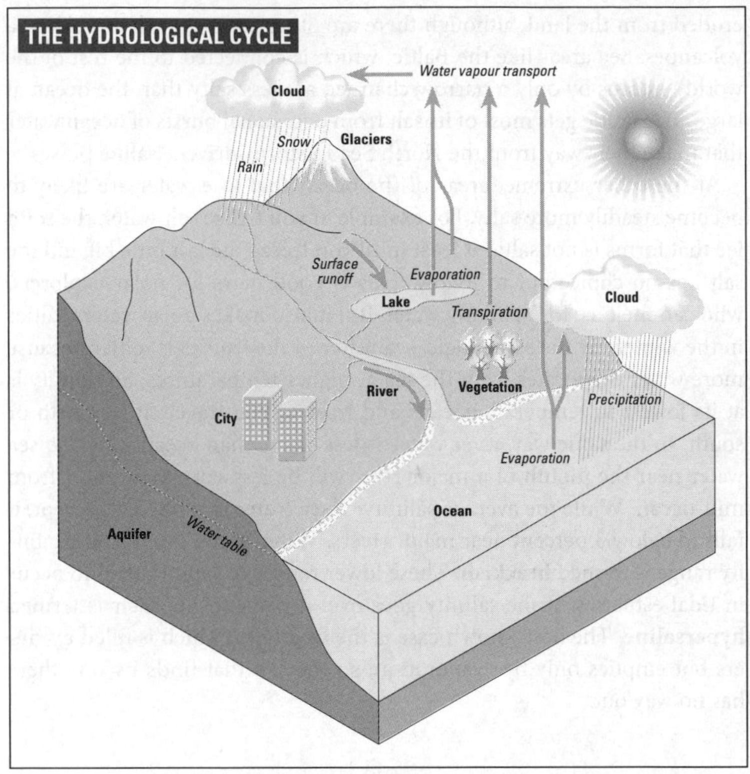

surface of a sea or ocean. Because it exists as both a liquid and a vapour at the Earth's surface temperature, water is constantly **evaporating** at the sea surface. As we saw in the previous chapter, this process is at its most vigorous in and around the tropics. As the air just above ground level is heated by surface warmth, it rises, taking water vapour with it. Clouds form and, at some point, the water in the clouds finds its way back to the surface as snow or rain.

Most of this evaporation (86 percent of it) comes from the oceans, but the majority of it ends up straight back there. Seventy-eight percent of **precipitation** also occurs over the oceans, which must make the hydrological cycle one of nature's less efficient machines. But the cycle gets more complex when rain or snow falls on land. If it falls as snow on Greenland or the Antarctic, water can be trapped, perhaps for millions of years, until the slow march of glaciers returns it to the ocean. But if it falls on a more temperate continent, any number of things can happen to it. Although its eventual fate is to close the cycle by returning to the oceans, water can spend thousands of years in this tangled part of the process. It might land in a **river** and head straight back to sea. Or it could fall on the land and flow over the surface to a river or stream, in which case it is classed as **runoff**. In the spring thaw, large quantities of **snowmelt** (runoff produced by melting snow) swell rivers and can lead to flooding. Rain might be absorbed into soil and end up in an underground water deposit, or **aquifer**. It could be absorbed by plant roots and taken into **living systems**, perhaps ending up inside a human body or that of some other animal. If you breathe a water molecule out, it becomes atmospheric water vapour once again.

Water on land is increasingly being captured for **human uses** such as drinking, agriculture, food manufacturing and industry. In the developed world the hydrological cycle on land is rarely seen working in its natural form. Even national parks and other preserved areas usually have their dams and flood-control systems.

Current affairs

Like the rocks beneath and the vapours above, the waters of the Earth see no reason to stay still. And if you recall the heat engine that drives the atmosphere from the previous chapter, you will soon see the analogies with ocean systems.

As with the atmosphere, the best place to start describing the way the oceans circulate is at the Equator, and on the surface. While the relatively stable environment in the deep ocean allows some variation in its composition, the top couple of hundred metres of the ocean are well mixed by the action of winds, tides and currents. This means the surface layer – or **mixed layer** – is much more homogeneous in its physical properties.

Because the ocean is heated from above by the Sun, the mixed layer is hotter than the rest of the ocean, often exceeding 20°C in the tropics. But below the mixed layer, at about 200m down, the temperature begins to fall steeply to about 4°C, in a layer known as the **thermocline**. At the same time, salinity rises from about 3.2 to 3.5 percent before more or less stabilizing below 1000m. This change is called the **halocline**. Although apparently tiny, it is significant because salinity, along with temperature, affects the density of sea water. As its salinity goes up and its temperature drops, the ocean water increases in density, mainly in a layer charmingly known as the **pycnocline**. The pycnocline zone extends down to about 1000m, although the exact depth varies from place to place and with the time of year, as the amount of solar energy arriving varies.

Because the water above the pycnocline is less dense than that below, it floats on top of it. The importance of this for the circulation of the oceans is that it allows the surface layer to sit on top of the deep ocean and obey its own rules.

Surface currents

The surface layer is home to the most obvious of the ocean's currents. They are largely driven by the wind and by Coriolis forces (see p.138), which are also influential in shaping the winds, as we have seen. So the best-known ocean current in the world, the **Gulf Stream**, follows the prevailing wind from the Gulf of Mexico to western Europe. By the time it reaches the coast of north-west Europe, where it is called the **North Atlantic Drift**, the water has cooled, and gained in salinity because it has lost water to evaporation and ice formation. It becomes so dense that it sinks and is returned to the Equator deep below the surface. However, some water also returns to the Equator on the surface, down the western coast of Europe and Africa, in the Canary Current, and back to the Gulf, via the Northern Equatorial Current.

Analogous **current gyres** – the equivalent of a cell in meteorology – are seen in the north Pacific, where the Kuroshio Current off Japan carries warm water north and the California Current brings it south again. The same picture is repeated in the southern hemisphere. Here the Brazil, Agulhas and East Australia Currents carry warm water southwards in the Atlantic, Indian and south Pacific Oceans respectively and the Benguela, West Australia and Peru Currents bring them back to the tropics.

Nearer the poles, things are a little less simple. In the southern hemisphere there is continuous ocean below Africa, Australia and South America in which both winds and waves flow non-stop from west to east, while smaller subpolar currents heading the opposite way hug the Antarctic coast. In the north, the **Beaufort Gyre**, a massive counter-clockwise current, slowly shifts ice as well as water around the pole, taking about four years to do a lap. It is named after the same Admiral Beaufort as the wind scale we met earlier. But the Arctic Ocean is not a closed system. It is connected to the rest of the world's oceans by the **Transpolar Drift**, which is fed by the great rivers of Siberia. It crosses the pole and ends up carrying water down the east coast of Greenland and into the Atlantic.

This is a very rough outline of the "**General Circulation of the Oceans**". But as usual with the Earth, things are not quite that simple. For a start, currents tend to be seasonal. The Gulf Stream shifts to the north in the northern hemisphere summer and south in the winter when the northern hemisphere cools. In addition, these currents affect the deep ocean as well as the surface layer. The Gulf Stream carries about 30 million cubic metres of water a second off Florida and much more (500 times as much water as the Amazon) by the time it reaches the mid-Atlantic. It is found at depth as well as at the surface. (Bizarre units corner: in oceanography circles, 30 million cubic metres a second is 30Sv or Sverdrup. The Sv, 1 million cubic metres a second, is named after the Norwegian explorer and oceanographer Harald Sverdrup.)

How fast is all this water moving? The fastest, topmost, part of the Gulf Stream can move at 2 metres a second, or 170km a day. This is one of the fastest major currents in the ocean, which means that the water it brings from the Gulf of Mexico is still warmer than the surrounding ocean water when it gets to Europe. Other major ocean currents tend to be slower, those at depth especially so.

The Conveyor

The deep currents of the oceans are driven in the first instance by gravitation, as water made denser by changes in temperature and salinity sinks through less dense water. The best-described example, as we saw above, is the descent to depth of cold, salty water off Europe after its journey from Florida in the Gulf Stream. The same effect is seen off the coasts of Japan and Antarctica.

The water that sinks in this way enters a machine known to science as the **Global Conveyor Belt**. The amounts of material that the conveyor belt shifts are on a planetary scale. Because it is primarily driven by differences in density determined by temperature and salinity, the Conveyor is also known as the **thermohaline circulation**. But there are other forces driving it too. The shape of the sea floor and the coastline will affect how currents move. And of course, our old friend M. Coriolis has to be involved – the Earth's rotation plays a key role.

Let's start where the Gulf Stream left off, where the now-cool water that carried heat from the tropics to western Europe plunges deep into the Atlantic off Norway. This water tracks back to the Equator along the coast of Canada and the US, but does not stop there. Instead it keeps going as a huge undersea river that flows further south, before branching in two. One branch runs round the bottom of Africa into the Indian Ocean, where it warms and rises. The other continues east around the Antarctic before being sucked north into the Pacific Ocean, almost to Alaska, where

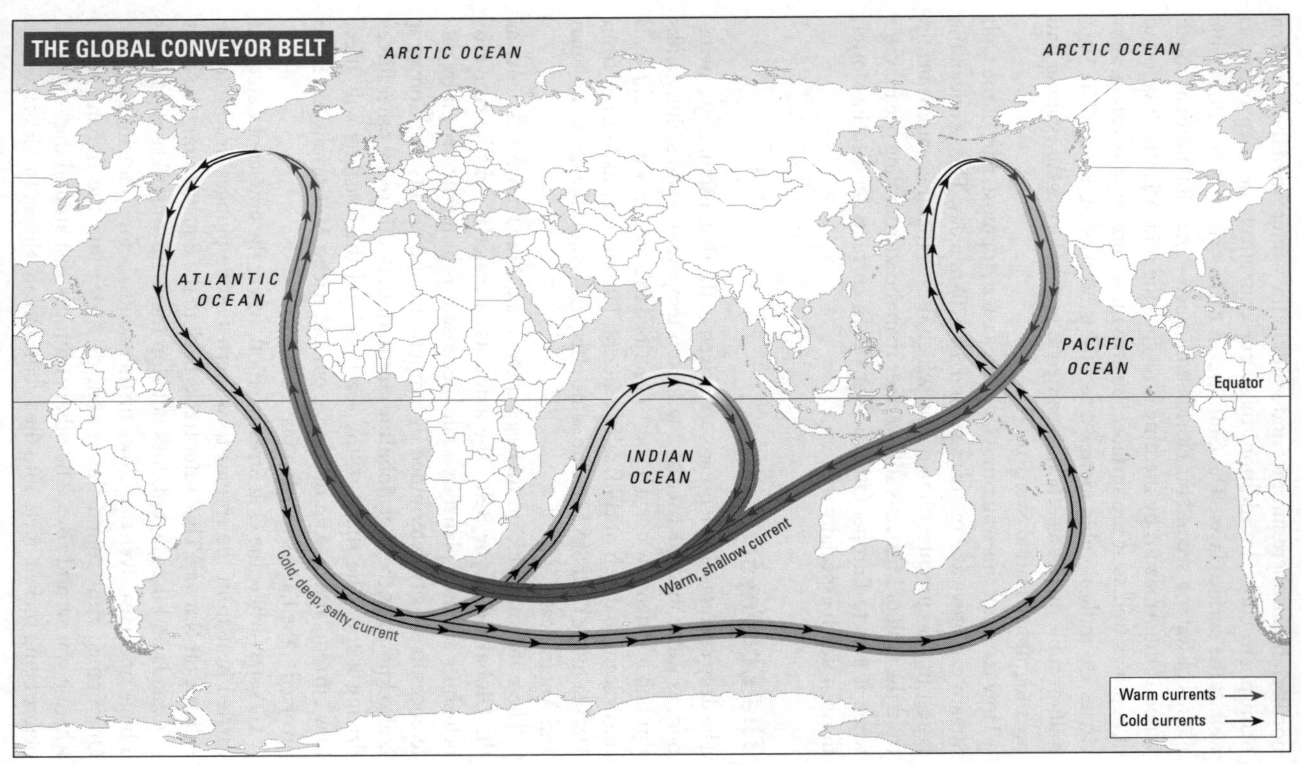
THE GLOBAL CONVEYOR BELT
ARCTIC OCEAN
ARCTIC OCEAN
ATLANTIC OCEAN
INDIAN OCEAN
PACIFIC OCEAN
Equator
Cold, deep, salty current
Warm, shallow current
Warm currents
Cold currents

it wells up and heads back to the Indian Ocean as a warm current. It's a truly global journey, but the conveyor is in no hurry: a complete cycle takes 1000 to 2000 years.

The details of this system are still not all known. Indeed, the term Global Conveyor Belt was only coined in the 1980s by Wallace Broecker of Columbia University in the US. But its importance is clear enough. It is the great world equalizer. Without it, the tropics would be hotter and the rest of the world colder. This applies especially to western Europe, which is an estimated 9°C warmer than it would be without the Gulf Stream. Have a look at the map. Newfoundland in Canada is about as far north as London – the one in England – but is a frozen waste, while southern England is temperate.

Our new knowledge of this temperature control machine has raised fears that human interference with the climate could change the workings of the Conveyor Belt. Certainly there seems to be less water moving through the Gulf Stream than in the past, although there is probably too little data to confirm any long-term trend. The idea is that global warming could melt the Arctic ice cap, flood the north Atlantic with light, low-salinity water and prevent that warm Gulf Stream water from keeping Europe in its current balmy state. The notion that humans could affect a system that carries far more energy than all the world's power stations put together may seem far-fetched, but it cannot be ruled out. More on that in the last chapter.

The sea, the sea

The deep and surface ocean currents carry far more than heat around the world. They mean that salt, oxygen and other chemicals are distributed more or less evenly around the Earth's oceans.

However, many of the Earth's most important water masses are seas rather than oceans. What they lack in size, they make up for in importance to humanity and to the Earth as a whole.

Just when a sea is a sea is a matter of judgement. The Red Sea, for example, is about as large as the piece of water variously known as the Gulf of Arabia or the Gulf of Iran. There is no good reason for one to be a sea and the other a gulf. Seas are more prone than oceans to political controversy, not least about their names. Iran and its Arab neighbours have bickered for decades over naming that stretch of water that cowardly cartographers call simply The Gulf. And what separates Korea from Japan? The East Sea for a Korean, but the Sea of Japan for the Japanese.

Wave power

Although currents both deep and shallow carry billions of tonnes of material around the Earth's seas and oceans, waves are the most visible manifestation of the power of the world's water.

Even a lake can contain worthwhile waves. But they are seen at their biggest in major oceans. The highest attested wave, seen during an enormous north Pacific storm by an observer on the US Navy oil tanker *Ramapo* in 1933, was agreed to be 34m high from crest to trough. This means that the wave would have 17,000kW of power for every metre of its length, so a 60m stretch would have the power of a large nuclear power station at full tilt. The biggest waves regularly observed occur where the flow of the full ocean is captured and funnelled, as with bores (see p.38), or when waves meet an oncoming current. This happens off South Africa where major ocean currents collide.

Waves form because as the wind blows across the sea, it rubs against the water surface and makes it slightly rough. As the surface gets rougher, there is steadily more for the wind to grab hold of and the waves get larger.

These waves involve a combination of transverse (up and down) and longitudinal (back and forth) motion, which means they have a movement in which the water particles perform a small circular orbit. This is why waves wash you back and forth a short distance when you are bathing in shallows at the beach. The water itself is not moving along the line of the wave, which is why boats bob up and down as a wave goes past, but are not washed ashore.

The scale of wave activity at any point in the ocean depends mainly on the amount of wind and the "fetch", the distance it has blown over the water. After a while, waves become "fully developed", which means that the same wind will not make them any bigger even if it blows across the water for a greater distance.

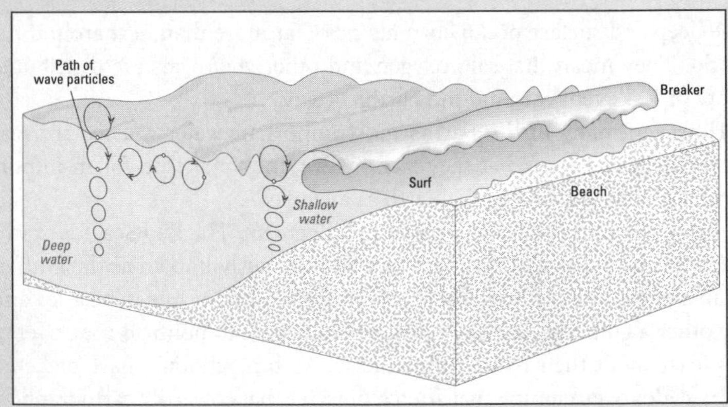

In the deep sea, all this works in an uncomplicated way because the base of the wave does not touch the seabed. But when they reach shallow water, and the invisible part of the wave interacts with the sea floor, the water is forced up and the wave gets higher. When it is one-seventh as high as the distance between waves, it becomes unstable, and collapses more or less rapidly. Then you have a **breaker**, the surfer's delight. Ideally, a surf beach should have a plentiful supply of ocean waves and a steep seabed that makes the breakers curl over dramatically.

Like light waves, water waves can be reflected and refracted. They can be seen being reflected off harbour walls, reducing the energy of the incoming waves and reducing the erosion they cause. In shallow water, their direction can be bent by refraction. The part that hits the shallows first slows down and the rest seems to bend as it catches up.

There are also waves deep below the surface of the ocean. These are longitudinal waves. Thinking back to p.102, they are like the P waves generated by earthquakes which can travel through the Earth's liquid core. In these waves, the water is compressed and then stretched along the line in which the wave is moving.

A breaker: placid out at sea, the wave collapses spectacularly as it reaches shallow water

Usually seas are comparatively shallow, existing on the continental shelf rather than the deep ocean. For example, the Yellow Sea between Korea and China sits on Asia's continental shelf and has a maximum depth of around 150m. However, those that are tectonically active, such as the Red Sea, the Mediterranean or the Caribbean, can be as deep as an ocean. The Caribbean is over 7000m deep in the Cayman Trench. To the purist, these are really micro-oceans rather than seas. Thus the East Sea/Sea of Japan is a deep sea that is effectively a detached piece of the Pacific Ocean.

However, a true sea has to be close to land while being more or less cut off from the full ocean, as with the North Sea, the Baltic or the Mediterranean. But it is probably wrong to think of a body of water as a sea unless it is attached to the rest of the world's oceans. So the Caspian Sea and the Sea of Aral are really inland lakes.

Their semi-detached status means that seas have their own ecology and way of working. The more isolated they are, the more individual they become over time. The ultimate case is the **Black Sea**. It has only a tenuous connection to the Mediterranean, which in turn has only a narrow passage to the Atlantic.

This means that its physical, chemical and biological nature is driven by the battle between the small amounts of salty water that make their way into it from the Mediterranean and the fresh water that arrives from the Don and the Danube, two of Europe's great rivers. This makes the Black Sea the world's biggest "meromictic" water body, one with a huge variation in salinity in a small region. The salty water builds up at the bottom of the sea and over time has formed a large volume of oxygen-free (or anoxic) water.

However, the connection to the Mediterranean is new. The Black Sea was isolated until about 7000 years ago, when it opened in a huge flood, memories of which may connect to the story of Noah. Until then it had many of its own fish, plants and other species, but now most of the Black Sea's flora and fauna have been replaced by Mediterranean types.

The **Mediterranean** itself shows that it is not just oceans that have a conveyor belt of their own in action. Its depth goes down to 5000m and it is tectonically active, but it is also being gradually squeezed as the African plate pushes up against Europe.

Until the Suez Canal opened in 1869 the Mediterranean had only one outlet, to the Atlantic at Gibraltar. Mediterranean water is saltier and denser than Atlantic water, so that it flows out below the surface at Gibraltar and is replaced by cooler water in a surface current. Like the global ocean conveyor, the Mediterranean conveyor is in no hurry. The whole cycle from Gibraltar to the east and back takes about seventy years. The escaping water is carried out into the Atlantic by the Coriolis Force. Some makes its way up the western coast of Europe and can bring Mediterranean species to the coast of Ireland.

By contrast the **North Sea** – the term took over from the previous German Ocean, oddly enough around the time of World War I – is more or less a piece of sunken land. It would not take much of an ice age to lower sea level enough to expose most of it to view. It rarely gets to 200m deep and on the fish-rich Dogger Bank, trawling for fish regularly brings up fossil trees instead. It takes in water from the Atlantic, the Baltic and the English Channel, and rivers such as the Thames and the Rhine. Off their estuaries the salinity can fall to below 3 percent. In the summer, the waters of the North Sea vary hugely in composition from one side to the other. But its winters are so rough that they mix its waters to a state of almost complete homogeneity.

The varying composition and origin of these waters is no bit of arcane theory. It drives the North Sea's ability to feed those who live near it. In the north, cold-water fish such as haddock abound, but further south, fishing yields warmer-water species such as bass. In addition, as recently flooded land, the North Sea has areas of sand, mud and gravel seabed, each preferred by different species. (See www.ices.dk/marineworld/fishmap/pdfs/factors.pdf for more information.)

But for the North Sea and most others, *Homo sapiens* is the species that matters most. Take a ship across it, and you will rarely be out of sight of another vessel. Its fishing grounds are the subject of dialogue-of-the-deaf arguments between scientists and the fishing industry over how badly damaged they are and how much fish they can afford to lose for human consumption. Gravel and other materials have been scooped from the seabed. The rocks that lie below it have been pumped for oil and gas, and in the era of renewable energy there are proposals to extract energy from it by the use of wind generators and machines to mine its tides and currents.

The confinement that gives different seas their individual nature means that they are prone to damage from over-use. They are easy to pollute and hard to clean up again. This same confinement can make them dangerous too. A stark example was seen in the North Sea in January 1953, when a **storm surge** swept clockwise round the United Kingdom. A storm surge occurs when low pressure causes winds to drive sea water towards the land. At their worst, this effect is supplemented by a high tide. This one began by sinking the ferry *Princess Victoria* between Scotland and Ireland, but became even more damaging in the North Sea where it combined with a very high tide. In the full ocean, such a surge could have worked off its energy without doing much damage. But in the confined and narrowing funnel that makes up the southern North Sea, the water could not escape and instead flooded low-lying areas of eastern England and the Netherlands. Over 1800 people were killed in the Netherlands, where 200,000 hectares of land were flooded, and 300 people in England. Waves over 6m high were observed on the normally placid coast of Lincolnshire in England.

Since this disaster, coastal defences have been improved in both countries, while computer systems to forecast such surges, and communications to get warnings to the people that need them, have been developed.

The confined and less stormy nature of seas means that they tend to have less craggy and forbidding coastlines than shores facing the full ocean. They tend to be the places to visit if you want to find beaches rather than cliffs, and if you prefer paddling to surfing.

Because there is more ocean than sea in the world, about 75 percent of the seafront on the planet is rocky and cliff-dominated. And as erosion is slower along the edge of a sea than of an ocean, seas, especially small ones, tend not to accumulate massive new quantities of sediment, unless there are rivers full of mud and sand to supply it. So it is at continental margins rather than in seas that new islands and deposits build up en masse.

However, there is a tricky balancing act at work on many sea coasts. Thus the east coast of England has large soft sandy stretches that are placid most of the time. But they are prone to sudden, violent erosion during severe storms that can remove many metres of material in a night.

Landlocked

The world's inland seas and lakes have a unique allure. Little ones such as those in the English Lake District or the Alps have had the power to inspire artists and writers from William Wordsworth and Mary Shelley onwards. Think of the wild weekend on the stormy shores of Lake Geneva in 1816 when the teenage Shelley penned *Frankenstein*.

But inland seas have other uses besides the artistic. For a start, they hold almost all the Earth's usable fresh water. Far more is contained in the ice caps of Greenland and Antarctica, but that water is locked up inaccessibly.

To count as a lake, a body of water must be disconnected from the world ocean. Instead, lakes are fed by rain and snow falling directly on them or arriving via rivers. They lose water by evaporation and via outflowing rivers.

Making a lake is simplicity itself. Start with a depression in the land, and water will find its way there, rising until it reaches the lowest part of the rim, where a stream or river will drain out. This origin makes lakes far less saline than seas and oceans. But they are also more confined and vulnerable. On the up side, this means that many lakes have unique ecologies. Like remote islands, many are home to species whose ancestors found their way in millennia before and have since evolved in response to local selection pressure. The down side is that a lake cannot rid itself of pollution as readily as an ocean might. It is surprisingly easy for large lakes to be damaged by human activity.

A prime case is the **Sea of Aral**, now divided between Uzbekistan and Kazakhstan but wrecked by the Soviet Union. The Soviet era's worship of industrial production at all costs led to the rivers feeding the sea, the Amu-Dar'ja and the Syr-Dar'ja, being used to irrigate cotton fields. The diverted water mainly evaporated or was lost due to bad sealing of the irrigation channels. But as a result of this water abstraction the Sea of Aral turned from the fourth-largest lake on Earth to the eighth. It shrank from 68,000 square kilometres in 1960 to under 29,000 in 1998 and the process is not complete yet. It could be down to 21,000 by the year 2010 according to DLR, the German space agency, whose scientists have modelled its decline. Its salinity has increased five-fold, making its water less

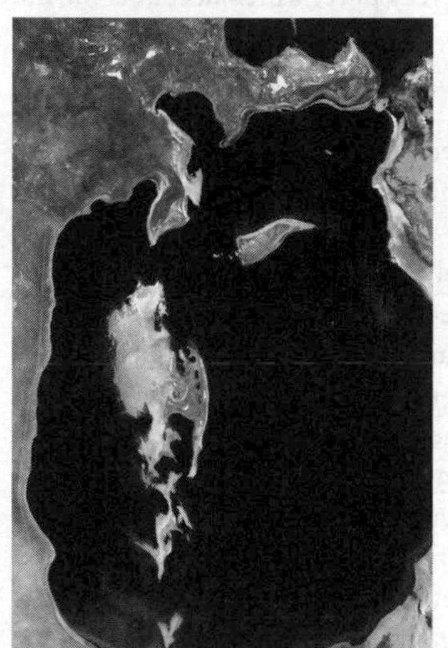

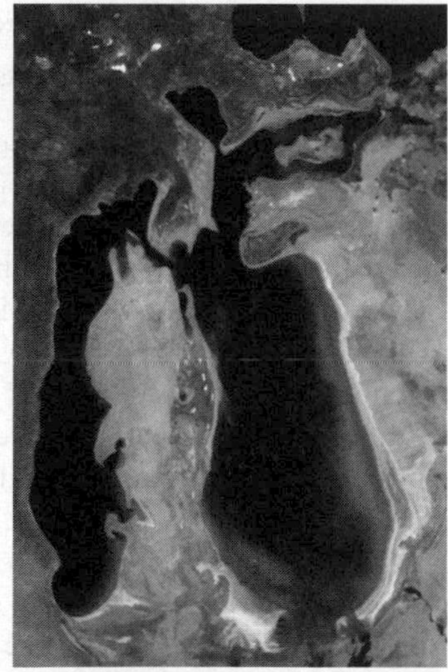

The Sea of Aral: vanishing between 1989 (left) and 2003 (right). There are plans to seal off the dry areas permanently to help the rest refill.

drinkable, while the total volume of water it contains has shrunk from over 1000 cubic kilometres to under 200.

In terms of economic activity, the decline of the Sea of Aral wrecked a fishing industry that used to produce over 40,000 tonnes of food a year. And as the lake has dried out, the salt it once contained has formed salt flats and blown across once-prosperous agricultural land. There are reports of increased respiratory disease and infant mortality among the surrounding human population to add to the depressing sight of big sea-going ships rusting in the sand, miles from any water.

For more information on the Sea of Aral, see:

Portrait of a Doomed Sea www.esa.int/esaEO/SEMUVZXO4HD_index_2.html

The Soviet Union was also home to one of the world's most remarkable bodies of fresh water, **Lake Baikal**. Despite some pollution from wood-pulp processing, its remote Siberian location protected it from severe environmental damage.

Lake Baikal has been an independent mass of fresh water for 25 million years. In this time it has developed its own ecology with unique species of everything from fish to bacteria. It contains about a fifth of the world's available fresh water, and runs down to over 1600m deep. Its watershed, the area that drains into it, covers 540,000 square kilometres. Most of it is rocky so the water entering the lake is very pure, which accounts for the lake's exceptionally clear water. See:

Lake Baikal Homepage www.irkutsk.org/baikal

In recent years, some answers have been emerging as to just why Lake Baikal has lasted so long in such a stable form. There had been a theory that the steep basin in which the lake sits is gradually subsiding. But a look at the bottom of the lake with a submersible shows that it is volcanically active and may be a site of slow tectonic spreading. The area's lively earthquake history supports this view, and the bottom of the lake exhibits volcanic vents like those seen at mid-ocean ridges. The lake may well be expanding at a rate of a few centimetres a year, a far-off effect of the deformation that has raised the Himalayas to the west. At this rate, even a 1km change in its size would take millennia.

Lake Baikal is not the only place where the unquiet Earth has generated stable geography. In East Africa, something similar is happening in the Rift Valley (see p.68). There Lake Tanganyika is the second-deepest lake on the planet, 1470m at its deepest. It has been isolated for about 10 million years. As a result it has about 600 unique species.

Nearby **Lake Victoria** can no longer make such a claim. Like other lakes in the area, it is home to unique species of fish from the cichlid family. There were 500 unique cichlid species in the lake, all descended from a common ancestor, until the 1950s when Nile perch were introduced to the lake in an attempt to improve fishing yields. These perch are able to grow up to 2m long, especially given a rich cichlid diet, and the species count is now nearer 250. Another Rift Valley lake, Lake Malawi, has fared better and is home to around 500 species. To see this in context, remember that Europe has about 60 species of freshwater fish in all its lakes and rivers put together. But at some point the spreading of the Rift Valley will mean that these lakes will meet up with other oceans and become salty, putting all these species in danger.

Tectonic processes do not have to produce vast expanses of deep water. Further north in the Rift Valley, in Kenya, the same forces have created shallow **soda lakes** (see box). Like the bigger lakes to their south, these lakes are far inland. But about 70 million years ago, modest water bodies

Soda lakes

Soda lakes are the favoured environment for flamingos, which – in the words of the Kenya Birds organization – enjoy standing in boiling caustic soda. They have a high concentration of dissolved alkalis, mainly sodium hydroxide. But how did they get that way?

Susan Baumgarte of Brunswick Technical University in Germany points out that such lakes are not unique to the Rift Valley of East Africa. Soda lakes and soda deserts can be found from Hungary to the US. They tend to be in places with unusual geology, where sodium and potassium dominate the metals dissolved in water. Both of these metals, unlike the more usual calcium or magnesium, produce soluble alkaline compounds that can be concentrated in water, as they are not removed by evaporation. They also occur most often in hot places and usually lose water only by evaporation, not through rivers. Some have sodium salts in such abundance that they can be mined industrially.

A soda lake's pH can be as high as 12.5 – in other words, they are very alkaline. But despite the hostility of this environment, they are brightly coloured with bacteria and other microfauna and flora. Some of these bacteria can even chew on denim, and have been used in the chemicals that create that pre-worn look in jeans. At the time of writing this was the subject of a legal action between the manufacturers and the Kenyan Wildlife Service.

Naturally, this abundant food supply is not going to go to waste. Hence the flamingos' enthusiasm for standing beak-down in such waters all day.

like these were probably present as the first symptom of the tectonic disturbance that marked the opening of the Atlantic.

However, big lakes can exist without volcanic turmoil, as evidenced by the **Great Lakes**, at the tectonically inert centre of North America. Put together, the five Great Lakes contain 23,000 cubic kilometres of water, about as much as Lake Baikal. The Great Lakes are at the centre of a continent, like Lake Baikal, in an area that exhibits the extreme weather variations that can be expected a long way from the ocean (see p.143). In turn, they are big enough, with an area of 244,000 square kilometres, to make their own weather. In winter, winds that blow across them pick up water vapour that reappears as snow far across the surrounding area.

The Great Lakes lose water into the St Lawrence River and other outlets only very slowly. They drain about 1 percent of their total volume each year. In the biggest, Lake Superior, the water is replaced only once in 191 years, but is very pure because it is comparatively far from cities and industry. In Lake Erie, the smallest of the five, the water is replaced much more quickly once every 2.6 years – but it is far more polluted by people, factories and agriculture. For more on the Great Lakes, see:

Great Lakes Atlas www.epa.gov/glnpo/atlas

Nor do lakes have to be on a spectacular scale to be rewarding. In the British Isles, the Alps, Scandinavia and other parts of Europe, many exist in mountain areas shaped by recent ice ages. An example is the Lake District in England. Its lakes are mainly in valleys carved by ice, which gives some of them a confined and sometimes gloomy appearance.

Here and in other mountain areas of the world, some of the most pleasing valleys have attracted the attention of water engineers who have improved on nature by constructing dams in areas where any other form of industrial development would be regarded as utterly inappropriate. Near major population centres (Manchester in the case of the Lake District, San Francisco for the Sierra Nevada of California), most of these dams and the reservoirs they retain are there to provide drinking water. In more remote areas, as in the Alps and Scandinavia, artificial lakes have been created, and existing ones expanded, to provide hydro-electricity.

Lakes do not have to be big to be important. Because the Earth's living systems are driven by water, even small ponds are vital to the ecology of both wet and dry areas. But big lakes have a very specific role in the Earth's water cycles. They aggregate water from many sources – Lake Baikal is said to have 544 tributaries. But they distribute it in concentrated form. In almost all cases they produce a single river, often a very large one such as the Nile, whose sources are in the great lakes of East Africa.

Old man River

Rivers, too, have their own hold on the human imagination. Most great cities that are not on the sea have a river running through them and would never have grown up without one. Nor is any painting of a wild landscape complete without flowing water.

At all points in its course, a river is best regarded as a transport system. Much as the amount of water it shifts varies over time, so does the quantity of other material such as sand, mud and rock that it can move. Winter, when there is most rain, and spring, when the snow melts, are peak times for river levels and for the erosion that goes with high water.

Rivers are at their most active when youthful. The mountains are the place to find the best waterfalls and rapids, which are the product of running water encountering harder and softer types of rock. In an era of rapid geological change, rivers are quick to change the landscape through which they run. This happened at the end of the last ice age, when new courses opened up for rivers and new valleys were cut. A symptom of mountain-building or rising land levels is the presence of "**incised meanders**". These are river bends in deep valleys, which have cut down as the land rose. One of the most spectacular surrounds the World Heritage Site at Durham, England (see photo overleaf). The sight of a piece of high land almost surrounded by a fast river was irresistible to medieval builders looking for a good defensive spot, so they enhanced nature by putting a magnificent castle and cathedral there.

Active rivers in high mountains erode the land through which they run both by water power and by the force of the rocks they carry with them. Sharp, fresh-looking rocks are characteristic of such rivers. Their banks are often marked by cracked and collapsing soil, protruding tree roots and other signs of rapid change. But eventually even the busiest rivers reach some sort of armed truce with the landscape around them, and settle between banks that are broken only rarely.

More mature rivers follow a different set of rules, appropriate to their role of crossing comparatively flat land areas. Over its whole 6270km length, the Mississippi falls by just 12cm per kilometre. However, even these serene-looking rivers change over time as they erode their banks and eat away softer material preferentially. As we saw in Chapter 3, the rule is that on the outside of a river bend, the water is at its fastest and most aggressive,

Incised meander at Durham, England

and erosion is at its most rapid. Speeds are slowest on the inside of the bend and material is more likely to be deposited there. Over time, this creates a series of wide bends, or **meanders**, in the river's course. As the river's path becomes more and more convoluted, the outside edges of two bends may eventually meet. Then the river cuts off the bend and straightens itself out, at least temporarily. The resulting cutout, called an **oxbow lake** (see colour section p.3), is one of the most distinctive structures on Earth.

The annual floods of the Nile, which provide thick, fertile sediment to the otherwise desert plains of Egypt, are proof that these apparently destructive natural processes are essential to life. There is always a fertile zone, the **riparian corridor**, along rivers even in the most hostile environments. But in the rich world especially, it is rare for mature rivers to be allowed to develop naturally. The land that they want to erode has owners who want to keep it. They argue for plenty of concrete to keep the river in its existing channel. By contrast, nobody has an interest in the new land that will emerge as the river shifts course. And while periodic flooding is natural, the people whose houses are in the way are quick to lobby for flood defence works to keep it in check.

Flood disasters

A combination of climate change, population growth and human migration to cities and river valleys means that the biggest flood disasters are probably ahead of us rather than in the past.

But the cheery folk at epicdisasters.com reckon that these are the biggest flood disasters ever. Readers in China may find this makes unpleasant reading.

Location	*Year*	*Death toll*
Yellow River, China	1931	1–3.7 million
Yellow River, China	1998	900,000–2,000,000
Yellow River, China	1938	500,000–900,000
Yellow River, China	1642	300,000
Ru River, China	1975	230,000
Yangtze River, China	1931	145,000
Holland/England	1099	100,000
Holland	1287	50,000
Neva River, Russia	1824	10,000
Holland	1421	10,000

These floods have widely varying causes. The Yellow River carries huge amounts of silt (hence the name), which build its banks up high. When they are eroded through, floods can cover vast areas rapidly. But the river's 1938 floods were caused deliberately to hold back Japanese invaders and the 1642 floods were created by rebels destroying the dykes. The Ru River disaster was caused by a dam collapse. Floods can be caused by water breaking through a defensive system, or on other occasions by its normal flow being blocked, for example by ice as in the Neva disaster near present-day St Petersburg.

Note that the 2005 inundation of New Orleans had far too low a death toll to figure here despite its immense financial and cultural cost.

When rivers reach the ocean

When rivers reach the sea, things become even more complex. A small stream can run into salt water without much ceremony. But a substantial river needs more, perhaps a delta or a full-scale estuary. A **river delta**, called after the triangular-shaped Greek letter delta, is essentially a land area with some water, while an **estuary** is more of a watery area with some land.

A delta builds up when a river deposits more sediment than the sea can erode away. Instead of one unique route, the flat topography of the deposits allows a number of solutions to the problem of getting water to the sea. The river cuts a number of distributaries which then start to meander so that some older parts become cut off. The definitive river delta is that of the Nile. But the delta formed by the mouths of the Danube as it enters the Black Sea in Romania is a more unspoilt example, as is the Lena River delta in Siberia (see colour section p.6).

Many river deltas occupy prime real estate – they are by the sea, on low-lying, fertile land. This means that the space they take up is in demand for farms, ports or factories. Often their untidy channels are straightened and deepened to allow ships to pass through. The mouth of the Rhône, on the Mediterranean coast of France, has been re-engineered with the addition of a new canal.

The basic equation of a delta or estuary is that when water slows to a halt, its ability to carry sediment drops to zero. This is why the Mississippi, which drains a vast area of the continental US, has generated a sediment area measuring 100km from north to south.

Even a river which carries too little sediment to build a delta can still carry enough material to silt up over time. This is especially true on a coast which is not being heavily eroded. In Britain, the west coast is rising and sediment is being deposited, so that major rivers such as the Mersey only stay open to shipping because they are dredged.

An estuary is a significant sheltered area of water where a river meets the sea. They are among the fastest-changing of all geographical features. Many, such as Chesapeake Bay on the Atlantic coast of the US, suffer severe erosion of their coasts and islands. The ecological value of estuaries lies in their basic property of mixing river and sea water. This means that they are home to many unique species of animals and plants which are adapted to changing salinity and to being exposed to air and water in turn several times a day. They are especially valuable as home to sea birds, shellfish including oysters, and seals and other marine mammals. Migratory birds often use them as stopover points on Earth-spanning flights.

The estuary of the Thames is a prime example. It stretches about 60km inland, with its inner reaches in London. Both banks are marked by a succession of marshes and areas of grassland. Even the wilder-looking parts are used for grazing cattle. The main distinction to be made between different areas of the seashore is whether they are marshland that floods at every tide – or perhaps every high tide – or flood plain, which gets submerged only when exceptional amounts of water are coming up or down the river, or doing both at once. Often the temptation to build on the flood plain is irresistible to developers. Flood defences can help buy time for their developments but the inevitable end is a damp one, and arrives all the more rapidly in an era of sea level rise.

For more on estuaries see:

US National Estuary Program www.epa.gov/owow/estuaries

A pattern of islands

Even the broadest sea is not a bare expanse of blue. If it were, where would treasure be hidden, and where would the castaways be cast?

Islands come in several flavours and we have met a few already. Some are simply accumulations of sediment that happen not to join onto the land. At major river outfalls such as the Bay of Bengal, new islands can arise and vanish in the timescale of human memory. The Bay sees the deposition of vast amounts of material but is also subject to severe storm surges that can remove it rapidly. Islands formed by such sediment tend to be long and narrow, with their long axis parallel to the flow of water. They are laid down at the point where the current carrying the sediment slowed.

But if the map shows you an island just offshore with a less regular shape, it is a fair bet that it is rockier and older. It is also likely to rise more imposingly from the sea. Rocky offshore islands are usually bits of fossil land that have not been eroded away yet. They tend to match the geology of the nearest piece of mainland. Limestone, sandstone and other hard rocks dominate their make-up and they form handy niches for seabirds and other hardy species. The prominent rock arches that are a feature of many sea coasts can be thought of as future islands, just as soon as the connecting rock has collapsed. But as the rocks in question are hard, the process could take thousands of years to show perceptible effects. Softer rock is removed faster but leaves bays rather than promontories.

An island found far out to sea is bound to have a different explanation. As we have seen, many are volcanic in origin, including some prominent island chains such as Hawaii, and big islands such as Iceland.

Such islands manage to break the ocean surface – and, in Hawaii's case, rise kilometres above it – because of the immense volumes of magma that the volcanoes that produce them generate. But not all subsea volcanism is so successful. It is now known that ocean floors, especially that of the Pacific, are strewn with thousands of "**seamounts**", underwater mountains defined as being more than 1000m high, measured from their local sea floor, but not reaching the sea surface.

It is common for new seamounts to be discovered. Many are in areas of intense seismic activity and are still live volcanoes, suggesting that they might accumulate enough material at some point to break the sea surface

New land

Those maps of the Earth look terribly solid, but don't be fooled. Just as Belarus or Eritrea can appear from something that was once labelled the Union of Soviet Socialist Republics or Ethiopia, new land can spring forth from the sea.

In a volcanic chain such as Hawaii, new islands are created first as lumps of lava at the sea floor. They then break the surface and stay there until they are eroded away again and become seamounts. But a look at the recent development of Iceland shows that nature does not always follow this stately process with a timescale of tens of thousands of years.

Take the island of **Surtsey**. It appeared off the south-west coast of Iceland, just beyond a long-established island called Heimaey, between 1963 and 1967. Now it rises to 174m above sea level, slightly less than when it first formed as the material that comprises it has compacted. Its total volume is about a cubic kilometre, of which 90 percent is below sea level. It is named after a Norse fire god.

Six years later, **Heimaey** itself was nearly destroyed as a human habitation by lava that threatened to cut off its harbour. It was saved by a lone enthusiast, and then the population at large, hosing the lava to cool it and stop it in its tracks.

Nor do new islands have to be natural. The 4 million tonnes of rubbish dumped in **Tokyo Bay** every year have already created about 250 hectares of new "land" there. There have been plans for a 30,000 hectare island in Tokyo Bay, to provide space and use up waste, although the Japanese economy would need more confidence than it has today (in 2006) for this to happen. Kobe, also in Japan, already has the artificial Port Island, so Tokyo may feel the need to compete.

as islands. Others have been above sea level in the past and have now been eroded back, as with the older members of the Hawaiian chain (see p.111).

Artwork depicting seamounts often has the vertical and horizontal scales fixed so that they look as steep as medieval towers. In fact, they slope at up to 25° but even these angles are not stable. Landslides on seamounts, perhaps set off by earthquakes, can cause **tsunamis**. Other seamounts can occur near land and can have sedimentary or volcanic origins.

Seamounts tend to end far enough below the sea surface to be of interest to scientists rather than sailors. **Reefs**, by contrast, are a byword for hazards to shipping.

They come in two sorts – coral and not coral – that have one thing in common. They both coincide almost exactly with the sea surface, allowing them to snare unwary vessels. If no coral is involved, a reef is simply a piece of rock, especially likely to appear on a coast which exhibits rapid erosion. Some are exposed at low tide. Around the world, many have been tamed – comparatively – by the addition of bells, lights and other warning devices.

A **coral reef** is equally dangerous to navigation, but is also one of the wonders of the world. A coral is a combination of alga and animal, and can only cope with shallow water, down to perhaps 30m, because the alga needs sunlight to carry out photosynthesis. Some corals are soft but others produce solid calcium carbonate that builds to a reef. In addition, other species such as sponges can secrete calcium carbonate, so coral should not take all the credit for reef-building.

Coral reefs can take a number of different forms. **Barrier reefs** are long reefs parallel to the shore that have formed in shallow water. They have a vital role in protecting the shore from erosion. An **atoll** is a ring of coral reef surrounding a lagoon. They are usually on the site of a sunken volcanic island. The atoll marks the initial coastline, and will remain only so long as coral growth keeps pace with the subsidence of the island.

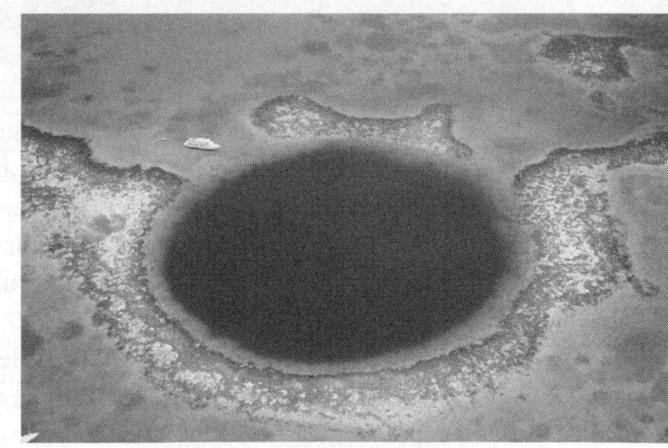

Reef Atoll Belize, surrounding the Blue Hole, a favourite diving spot

Geosight #4: The Great Barrier Reef

Its Australian publicists term the Great Barrier Reef "the largest natural feature on Earth", which it is not. There are plenty of mountain ranges that beat it for size. But their enthusiasm is understandable. The reef (see colour section p.6) is undoubtedly the largest living thing on Earth, running for 2300km off the north-east coast of Australia. Its northward boundary seems to be the point at which the water gets too hot for the corals that build it to survive.

Its physical foundation is the work of 359 species of hard corals that manufacture limestone and give the reef its solid structure. They are accompanied by soft corals. These are far more hazardous to humans than hard ones, because they have developed toxins to deter predators. Indeed, the whole reef is a chemical warfare zone. The jellyfish have vicious stings, and one called the Box Jellyfish is said to be the most hazardous animal in the world to humans. Even the fish are not to be messed with. The stonefish gets its name from its rocky camouflage but also has deadly venomous spines.

However, none of these defences are effective against the challenges that humanity offers the reef. The sheer number of people that go there means they are a pollution hazard, even if all they do is perform normal human bodily functions during their visit. They tend to arrive in boats, whose anchors can damage the coral.

The reef now has management who aim to minimize such local hazards. But in the longer term, bigger environmental changes could be harder to avoid. If sea level rises, can coral grow fast enough to keep up? In the past, the answer has been yes, but it might not stay that way. Or might hotter seas reduce the range of the reef? Hot water can kill by "coral bleaching" in which the heat kills the algae that sustain it.

For more information see:

Great Barrier Reef Marine Park Authority www.gbrmpa.gov.au

Coral-bearing limestone is known from hundreds of millions of years ago. But it has been in a conspicuous golden age of late. As the last ice age ended and sea levels rose, corals grew to keep pace with the rise in water levels by building higher on the limestone laid down by their predecessors. Oceanographers now use coral as a ruler for measuring sea level change.

In the modern era, however, the picture is less favourable. As well as sea level rise, global warming causes sea temperatures to rise. Corals find it hard to cope with warmer water, which is why some of the world's hottest ocean waters, for example in the Arabian Gulf, do not have coral. In 1998 and 2004, when the El Niño effect (see box overleaf) caused sea temperatures to rise well above normal, many tropical coral reefs suffered from "coral bleaching" or were killed. If human-induced climate change causes sea temperatures to rise in the future, more coral will die. In addition, many reefs are subject to human-induced stress at the local level. Some are mined for rock, built on by hotel owners, or polluted. As they can only exist in shallow water and are usually near to land, they are often in places where pollution disperses only slowly.

Unseen seas

The reserve of water that is thought about least often is the one you cannot see. Billions of tonnes of the stuff exists just below our feet and has a vital role in the Earth's living and non-living systems.

As we have seen, the Earth gets steadily hotter with depth, and soon reaches the boiling point of water at sea level, 100°C. But as you get deeper, the pressure grows as well, and higher pressure means a higher boiling point. This allows water to exist in the Earth's crust at temperatures above 100°C.

But this cannot go on for ever. Water has a "critical point" at 374°C beyond which it cannot stay liquid and is always a vapour. This means that water only exists inside the Earth at comparatively modest depths. There are no secret underground lakes many kilometres below your feet.

Many of the solid minerals that make up rocks contain water as a part of their chemical composition. This water is locked away stably in their crystals and is inaccessible unless you take the rock and heat it to destruction. But much of the water within the Earth is "available" water that is mobile and active.

The interior of the Earth is as full a participant in the hydrological cycle as the seas, clouds and rivers. But when you dig into the ground, you don't usually find water, at least not to begin with. The key concept here is the **water table**.

The water table is the top of the underground zone that is saturated in water. This zone is called the **aquifer**, a term that comes from the Latin for water-bearing. But unlike most tables, the water table is not a flat, stable structure. It rises and falls roughly in sympathy with the land surface above it. Otherwise everyone who lives in a valley would be paddling while mountain-dwellers would need to drill wells thousands of metres

Niño or Niña?

El Niño has hit the headlines as a codeword for bad weather, especially in the Pacific and California. But there is a little more to it than that.

The name El Niño might suggest a relatively benign phenomenon. Would the devout Catholics of Latin America call something after the Baby Jesus if it was particularly evil? However, although the phenomenon is usually observed around Christmas time – hence the name – it is a less welcome arrival. The term was first used by fishermen to describe a warm sea current running up the Pacific coast of South America and Mexico at this time of year. It happens when the easterly winds that usually drive water across the south Pacific weaken and allow warmer water to intrude. El Niños tend to occur every three to seven years.

La Niña is the opposite effect and happens when colder-than-usual water takes over the eastern Pacific. It is about half as frequent as El Niño, and less damaging.

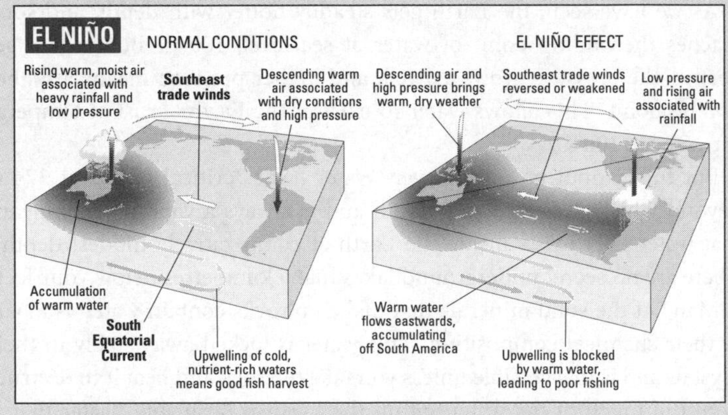

deep to get a drink. You could regard a lake or marsh as a place where the water table breaks out above ground level.

But it would be wrong to think of the zone below the water table as a kind of subterranean water tank. It consists of rock or, if the water table is closer to the surface, unconsolidated sediment, with water in the pores between the grains. Some rocks are better than others at soaking up fluids. Limestones and sandstones are better than most other sedimentary rocks, and sedimentary rocks are preferable to igneous ones. People hunting for large bodies of underground water look for such permeable rocks, as do prospectors for oil and gas, which occupy the same intergranular pores as water.

Although El Niño is a warm current that heats the land, it is unwelcome because the water it carries is poor in nutrients. Without the easterly winds, there is no force to drag nutrient-rich water from depth to feed the fish on which local economies depend. But the real importance of El Niño has only become apparent in recent years. It runs far beyond the fish markets of Peru.

El Niño is associated with higher-than-usual rainfall in the southern US. In 1997–98, an especially severe El Niño brought fatal flash floods to California. By contrast, La Niña years are dry in the southern US but wetter to the north.

However, El Niño also affects societies which are in less robust condition to fight back. For example, it is associated with droughts in Indonesia, caused by the absence of the cross-Pacific wind that should bring the rain clouds. Droughts associated with El Niño have also been observed in Australia, the Philippines, Brazil and southern Africa. The weather system affected is called the Southern Oscillation, and the whole combination of effects is called **ENSO** for El Niño/Southern Oscillation.

One topic on which there is no agreement is whether **global warming** might make El Niño worse in the future. Some computer models think it will; some say it will make future El Niños less intense; some say the difference will be slight. But by making bad weather more news-worthy, global warming has certainly made El Niño more famous.

For more information, see:

University of Illinois El Niño Guide
ww2010.atmos.uiuc.edu/(gh)/guides/mtr/eln/home.rxml

NOAA El Niño Page www.elnino.noaa.gov

Water gets into the aquifer via "recharge" areas. Often these are wooded or vegetated regions. Water falling on rocky hillsides tends to be carried off into streams and misses joining the aquifer. Human development of land into cities makes things worse. When a site is developed, surface water is usually captured and sent into rivers down concrete channels and pipes. This cuts down the amount of water getting to the water table.

Both this pressure and the growing human demand for water means that the water table is on the retreat in many parts of the world. At the time of writing, some of the most dramatic such effects are being seen in China. They result from the country's rapid economic growth and its population's increased enthusiasm for Western-style food and drink, which require more water-intensive crops.

However, the water table can rise as well as fall. People and industry in London used local wells as their water supply for many centuries. Now they get it from far-distant sources, and little water-intensive manufacturing goes on in the city centre. As a result the water table is now rising back to the level it must have occupied in pre-industrial times, and threatening to dampen deep basements as it does so.

In less habitable parts of the world, underground water takes on an even more vital importance. The Sahara Desert covers nearly 9 million square kilometres and – surprisingly – supports a population of 2.5 million people. They congregate in places where water can be found. The oases that make life possible there are almost all in low-lying areas where the water table is closest to the surface.

Underground rivers

The water in them is replenished by **underground rivers** such as those that flow from the Atlas mountains to the west. And in Egypt, new underground rivers have been detected by satellite sensors like those used to hunt for water on Mars.

To get truly majestic underground lakes and rivers takes some specific geology. Because the Earth is basically a solid body, you need some soluble rock. One candidate is **salt**, sodium chloride. It exists in the Earth's crust in domes, needles and other shapes, and is extracted for human use by being dissolved in water and pumped out. Human salt mining has created spectacular caves in many parts of the world, notably the World Heritage Site at Wieliczka near Krakow in Poland. Caves exist in Spain and Israel that have been created in this way by nature, but they are a rarity.

Most of the world's natural caves were created by water, but using a different process. As we saw in Chapter 3, because the atmosphere contains carbon dioxide, which is slightly soluble in water, rain and river water do too. Let it loose on **limestone**, which is more or less pure calcium carbonate, and a slow chemical reaction will produce calcium bicarbonate, which is soluble in water.

This means that anywhere in the world where you find limestone, you find caves. Because the caves are formed by water, they are essentially underground stream courses. Like streams on the surface, the water they contain is quick to rise when it rains hard. Many of the worst caving disasters happen when people are trapped by rising water.

However, many caves are now dry and completely safe. They get this way when a cave has been cut by water which has then dissolved an even lower passageway and left the upper one dry. To the true caver, a cave system has all the pleasures of a river system on the surface, but with the added attraction of being in three dimensions.

But even people who like to stay on the surface can learn a lot from the way water acts underground. Because it can move through rocks such as sandstone and limestone, springs and streams tend to appear where these rocks meet up with something less permeable such as shale. And a river vanishing into the ground is more or less a certain indicator of limestone, unless human engineers have been at work. So you can glance down a valley and draw your own mental map of its geology as you see water come and go. Along some valleys in Europe, you can map the underground world by looking at the villages – each marks a geological join which forced water to appear above ground and made human habitation possible.

The ultimate mineral

There are now 6.5 billion people living on the Earth. They use about 26 trillion tonnes of water a year for everything from irrigation and drinking to steel-making and washing their cars. By contrast, they use "only" about 4 billion tonnes of oil a year. So, although most water isn't traded, it is a more significant commodity than oil even if the sums of money for which it changes hands are smaller. This means water is a mineral in two senses. It makes up a goodly percentage of the Earth's crust and takes part in Earth's natural processes, and it has immense value to people, making it an "**economic mineral**" as well as a geological one.

How much water are people using? Like every other resource, rich people get through more of the stuff than the poor. In Africa, each person uses on average 254 cubic metres a year, but in Western Europe, the figure is 1280 cubic metres. In South America, it is 478 cubic metres, and in North and Central America, 1881 cubic metres. Asia and the former USSR are somewhere in between at 519 and 713 cubic metres respectively. (These figures and some of what follows comes with thanks from data gathered by the University of Michigan.)

A clearer way to think of it is to consider how much of the world's accessible water we are using. As we have seen, most water on the Earth is too salty to drink and most of the rest is inaccessible. But even this understates the problem. Huge amounts of fresh water flow along the

Amazon but there is almost nobody there to drink it. The same applies to the big rivers that flow into the Arctic. By contrast, the Colorado River in the US is so heavily used that its mouth has dried up. In most years, no water arrives there.

It seems that all the world's plants and forests get through about 70,000 cubic kilometres of water per year, and that agriculture and forestry account for about a quarter of this.

At the same time, the rivers and other watercourses of the world drain about 41,000 cubic kilometres a year of water. But about 70 percent of it is inaccessible except to ecotourists and scattered human populations. Of the remaining 30 percent, over half is being captured for human use, according to Sandra Postel of the Global Water Policy Project. She adds that about two-thirds of this water is used to irrigate crops. Changing food tastes and population growth will increase the pressure to use more water in this way. More dams might add a little to the available supply but as we have seen, the hydrological cycle is a closed system where the total amount of water does not change.

Water wars

In some parts of the world, water is already politically hot, and nowhere more so than in the **Middle East**. It undoubtedly plays a major role in the conflict between Israel and its neighbours, including Jordan and Palestine. A substantial aquifer lies below the west bank of the Jordan which Israel has been accused of depleting. Certainly people in Israel, both Jews and Arabs, use more water, mainly for agriculture, than their neighbours. Israeli exports include fruit and cotton, both water-intensive crops. Irrigation for cotton-growing, as we have seen, has already been responsible for the destruction of the Sea of Aral. Recent reports suggest that water levels in the Sea of Galilee are heading the same way. If the water levels fall to a critical level, lower-lying water that is dangerously saline may be drawn to the surface and damage the lake for fishing, water supply and other uses.

The aquifer in question is one of the most extraordinary known to geology. While the water table normally follows the ground level, this one dives from 400m above sea level to 400m below it in the space of 30km, between the heights near Jerusalem and the depths of the Dead Sea. Part of the problem is that traditional agriculture in the area depended on the "upper aquifer" of the Judea area. This aquifer is separated by an **aquitard** – a band of impermeable rock – from the much larger lower aquifer below it.

The upper aquifer has sufficed for millennia to keep crops watered and people refreshed. At a reasonable rate of use, it can be replenished by rainwater and melting snow. The lower one can only be accessed by high-technology drilling. It contains ancient "fossil" water that can only be replaced over geological time. Using it up is like depleting an oil reserve.

The same pressures are driving squabbles over other great rivers. The names of the **Tigris** and the **Euphrates** are among the most resonant of any rivers. They were the great watercourses of ancient Babylon and now of Iraq. But they also water Turkey and Syria, and are important to the Kurdish people who live across the region's frontiers.

There has been a persistent failure to agree on the division of the precious water of these rivers. Turkey has announced plans for the **Great Anatolian Project**, involving 22 dams and 19 hydroelectric power stations on their upper reaches, which would have damaging effects on the water flow lower down.

At the mouths of the Tigris and Euphrates is another long-running political crisis, concerning the future of a large area of wetland and of the Marsh Arabs who live there. The former regime of Saddam Hussein used drainage as a form of environmental warfare against the Marsh Arabs,

The Marsh Arabs as seen in 1974. The value of their culture and habitat is now more likely to be appreciated than it was in the past.

building dams on both rivers to reduce the water flow to the marshes. But even in their previous form, the marshes were artificial, depending upon human maintenance of their drainage systems. Despite efforts to restore the marshes, less than 5 percent are left today, and most of the Marsh Arabs have been forced to leave.

The idea of "reclaiming" the marshes was in the minds of politicians long before Saddam came along. It was developed by a British civil engineer, Frank Haigh, in a report to the Iraqi government in 1951. His intention was to capture the water in the marshes, and the land they occupy, for agricultural production, in the era before such an environment was regarded as worth protecting in its own right. (For more information see www.american.edu/ted/ice/marsh.htm.)

Many of the same debates surround the **Nile**. Although everyone knows that Egypt is "the gift of the Nile", the river actually flows through ten countries, as befits the world's longest watercourse. Its use is governed by a 1929 treaty. Even Britain signed it, as a major colonial power of the time.

The treaty has come under pressure because it essentially prevents anyone upstream of Egypt from using any significant amount of its water. People as far away as Tanzania are supposed to get Egyptian consent to use Nile water to irrigate their fields. The discussions have been marked by many threats to withdraw from the treaty. But if it is abolished, the agreement that replaces it will need to be carefully crafted. Demand for water is growing all along the Nile and it is possible, as we shall see later, that the basic shape of the Nile system will be altered by climate change.

These are only a few of the world's water conflicts. Others affect the **Ganges** – between India and Bangladesh – and the **Mekong**, involving Vietnam, Thailand, Laos and Cambodia. But the highest-profile water disputes are all going on in the Middle East, and there are two reasons for this. One is that the area is an arid one where population growth, changing agricultural practices and other pressures are leading to growing demands for a scarce resource. The problem may be eased if new technology can reduce demand for water, or make more appear by producing affordable methods of desalinating sea water.

But perhaps the real issue is that an area that has poor political institutions which are unable to solve international, or indeed internal, disputes lacks the machinery to resolve an issue such as water. Rivers go where they want and have no interest in boundaries. But in Europe, rivers such as the Rhine wander from country to country, or form international frontiers, with little controversy. Indeed, their management often provides a good example of international co-operation. Nor do Canada and the United States rattle their missiles to assert rights over the Great Lakes.

Perhaps the political activists and journalists who like to talk about "water wars" have missed the point. Water is routinely used as a weapon. This tactic is threatening because it attacks people, their food supplies and their entire environment at the same time. But an area that has proper governments and working diplomacy can solve these problems without conflict. The Worldwatch Institute in Washington DC has found that water disputes are usually resolved peacefully. Perhaps fittingly for the world's most important commodity, says the Institute, water negotiations are often among the most successful and conflict-free of human disputes.

For more information on conflicts over water see:

Global Policy Forum www.globalpolicy.org/security/natres/waterindex.htm

GDRC Water Crisis page www.gdrc.org/uem/water/water-crisis.html

7

The icy Earth

The icy Earth

The Earth's frozen areas are many. They currently cover about 10 percent of its surface. At the height of the most recent ice age the figure was about 32 percent. In this chapter we will explore the great icy zones of the planet, before looking into the Earth's icier past and the current threats to the Earth's ice cover.

The Antarctic

In the icy world, nothing can compare in size or importance to the Antarctic, which is almost a complete cryo-continent, covering about 14 million square kilometres, about as much as Mexico and the continental US put together. It has been a separate continent in the region of the South Pole for over 50 million years and has the most hostile environment of any area of the Earth's surface.

If you recall Chapter 5, you already know part of the problem. In the Arctic and Antarctic, winds start out at the poles and spread outwards. The air that makes them up starts out dry. In the Antarctic, it stays that way because it is blowing over land. So the Antarctic is a desert. Most of it has precipitation that would add up to less than 50mm a year if it fell as rain, which of course it usually does not.

The Sun never gets far above the Antarctic horizon, and it vanishes completely for up to six months of the year. In addition, the Antarctic runs up to over 4000m above sea level. This makes it even colder. Finally, because 97 percent of the Antarctic continent is covered in ice and snow, it reflects away most of the solar energy that does manage to arrive. The albedo (reflectivity) of snow can be up to 0.85, whereas the albedo of the Earth at large, as we saw in Chapter 1, is only about 0.37. All this means that the Antarctic has an average temperature of -49°C.

The ice sheets of the Antarctic are tricky to measure. But in recent years, observers have converged on a figure of about 30 million cubic kilometres for the amount of ice they contain. The perfectionists at the US Geological Survey plump for 30.1098 million, which is taking accuracy too far. This means, as one enthusiast put it, that chopping it up would give everyone on Earth a piece of ice the size of the Great Pyramid at Giza. More significant than this improbable feat is that if all this ice melted, sea level would rise by about 70m. It also adds up to 70 percent of the world's fresh water and 90 percent of its ice.

Near the edge of the Antarctic continent, ice sheets are grounded on rock below sea level. But further out, they float and are called ice shelves. These advance and retreat in winter and summer respectively and are a source of icebergs. The main body of the continent is divided into the East and West Antarctic Ice Sheets. The West Sheet is the one you tend to hear about in news reports. This is because large parts of it rest on rock

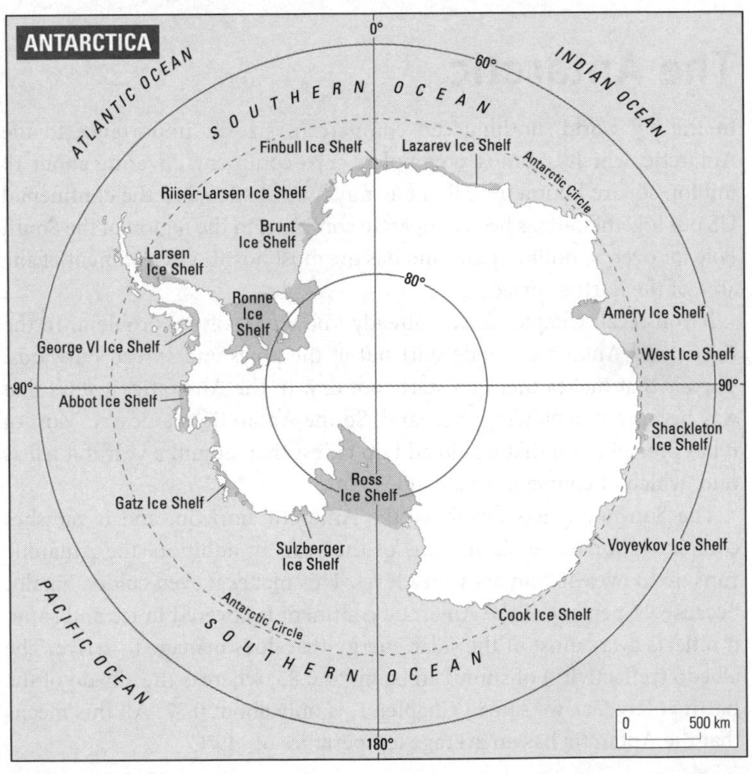

below sea level. If the Earth warms, it is especially likely to melt or break up, with serious effects on sea level. However, this applies to parts of the East Sheet too. In addition, there is the Antarctic Peninsula, pointing in the direction of South America.

It may be the coldest and least inviting continent, but how many others change their shape completely over the year as Antarctica does? Between summer and winter, it grows in size as the sea freezes. In the height of summer, its land area is supplemented by up to 3 million square kilometres of **ice shelf**, with the Ross Ice Shelf (see colour section p.7) the biggest. But in the winter, this area can grow by another 15 million square kilometres.

As well as all this ice, the region around the Antarctic continent contains an abundance of islands. Many were used for whaling stations in the past and are now home to scientific bases. The most astonishing is **Ross Island** in the Ross Sea. Although most of the Antarctic is tectonically quiet, it does have some volcanic and seismic activity. On Ross Island it breaks out in the form of **Mount Erebus**, a full-scale volcano rising to 3800m above sea level. Erebus has been continuously active for decades and, inevitably, there is now a webcam for you to have a look.

Mount Erebus Volcano Observatory www.ees.nmt.edu/Geop/mevo/mevo

On the coasts and islands of Antarctica the heights above sea level are lower, there is sea to moderate the climate, and the Sun gets that much higher in the sky. So these parts are more temperate and are home to more life, including all of the continent's bigger life forms. They include birds such as the skua and big sea mammals such as seals that are known from elsewhere in the world, as well as penguins, unique fishing and swimming birds. The oceans off the continent are also full of fish and other life ranging in size from krill – shrimp-like crustaceans – up to walruses, sea-lions and whales.

The inland parts of Antarctica are not one blank and unvarying sheet of ice, however. In 1903, the British explorer Robert Falcon Scott discovered the Taylor Valley, the first of the Antarctic **dry valleys** to be found, and called it "the valley of the dead". The dry valleys are areas where the wind scours away the snow to leave bare rock and sand, and they are the most significant ice-free areas of the continent. But even the dry valleys are less barren than Scott thought. We now know that they are home to nematode worms and that there is a simple but effective ecology of bacteria in the valleys' permanently ice-covered lakes. The valleys are the nearest thing the Earth has to the conditions today on the surface of Mars. The fact that life exists there is regarded as an indicator that it could survive on Mars.

Icebergs ahoy!

The thing that everyone knows about icebergs is that about 90 percent of the things are below water. Unusually for such a widely held belief, this one is completely true.

Archimedes lived in Sicily and it is safe to say he never saw an iceberg, but he could have told you why. His Principle tells you that anything placed in water is pushed upwards by a force equal to the weight of the water it has displaced. Ice is about 10 percent less dense than water. So when 90 percent of it is under water, the force of the water pushing upwards matches the weight of the ice pushing down, and it floats with about 10 percent of its mass above the surface. However, a fresh iceberg often has a less dense snow layer on top so that more than 10 percent of the volume is visible.

The icebergs that float in the world's oceans start out in the Arctic and Antarctic. They have two main sources. Some break off from the seaward end of a glacier, while others are formed when part of an ice sheet breaks loose. Either way, an iceberg is always "**calved**" rather than just forming. Not surprisingly, iceberg formation is at its height in spring and summer as the ice that formed over winter breaks loose. The ***Titanic*** was sunk in April 1912, at the height of the season for icebergs heading into the Atlantic from the Canadian and Greenland Arctic.

Because of the sheer scale of the Antarctic, and the fact that it is a frozen continent completely surrounded by ocean, it produces most of the biggest icebergs that enter the world's oceans. Fortunately, the Southern Ocean is so

Iceberg calving at Paradise Bay, Antarctica

empty that they are not too big a hazard to shipping. Icebergs here tend to be kilometres rather than hundreds of metres in size.

According to the US government's National Ice Agency, the ice sheets of Greenland and Canada produce between 10,000 and 40,000 bergs per year, of which about 375 make their way into the North Atlantic shipping lanes. Most come from West Greenland, with smaller numbers from East Greenland and the Canadian Arctic. They take their time, typically three years, to get from calving to sailing the open Atlantic.

The smallest icebergs are called **growlers** – less than 20 square metres – and **bergy bits**, which are broken off bits of glacier ice a few hundred square metres in size. Even these will wreck your yacht if you hit one. Bigger ones include "**tabular**" icebergs, the colossal aircraft-carrier-shaped ones. A **floe** is a large floating piece of ice that is more or less flat.

Tabular icebergs cruising the Southern Ocean

Icebergs are so huge below the water that their scrape-marks can be seen on rocks and the ocean floor. Louis Agassiz's sighting of such marks (see pp.219–20) merits a mention by Herman Melville in *Moby Dick*. Their immense bulk means that icebergs are a hazard to ships and even more to fixed structures such as oil rigs. Sometimes ships are used to tow away especially threatening bergs that approach an oil platform. Nowadays, land-based and satellite radar can be used to track icebergs, much reducing their threat.

So far, nobody has tried the oft-discussed idea of towing Antarctic or Atlantic icebergs to water-deprived areas of the world such as the Middle East or California. Calculations suggest that with a fast enough ship, it might be possible to get them there before they melt. Much debated in the 1970s, this idea now seems to have been abandoned.

But there is a world of difference between life hanging on in the dry valleys and life originating there. The life we see there is a highly adapted version of living systems that got their start in more temperate zones of the Earth.

Even more astounding has been the discovery of over seventy liquid water lakes far below the ice surface of the Antarctic. With the Russians still in pole position in Antarctic research, it is no surprise that the biggest such lake is called **Vostok**, Russian for east. It was found in 1996, partly by observations of the surface above. As the ice above Vostok is floating on the water below, it is noticeably flatter than the rest of the local landscape.

Lake Vostok lies below 4000m of ice, but stays liquid because of heat from the Earth below and pressure from above. It is about 250km long, 40km wide and about 400m deep, or, as NASA puts it, about the size of Lake Ontario. It seems to be in an area of tectonic activity, upping the amount of heat it receives from below, a little like an under-ice version of Lake Baikal in Siberia (see p.182). But it has been hermetically sealed off from the rest of the world even more effectively than Baikal and for even longer, perhaps for more than a million years.

This means that Vostok is one of the most precious conservation assets on Earth. Whether it is possible to take a look without introducing organisms from the outside and wrecking it remains to be seen. So far, investigative drilling has stopped a couple of hundred metres short of the lake. But evidence has been found there that suggests the lake may be home to microbial forms of life.

Even the thick ice that covers most of the continent is far from dead. It looks completely barren but the ice cores drilled out of it contain abundant bacteria.

The world's largest lab

The abundance of life, especially krill and whales, in the oceans around Antarctica has long attracted the human hunter. And in recent years increasing numbers of tourist ships have been visiting the region. But plans to exploit minerals in the Antarctic oceans and continent have failed to take off for commercial reasons and because of environmental concerns. The result is that the Antarctic has been left almost entirely to the scientists, and has become the world's biggest science lab.

As we have seen, the Antarctic ice is about 4500m deep at its thickest point. This means that it has taken time to build up. In the process, it has turned into an invaluable record of the Earth's past climate. **Ice cores** drilled in the Antarctic by Russian scientists have produced a 200,000-year-plus account of the temperature of the Earth, using the varying composition of the ice as a thermometer (see box on pp.214–15).

In addition, up on the central plateau, astronomers use telescopes and cosmic ray detection devices to probe the outer reaches of the universe. On the peninsula and the islands, as well as in the dry valleys and the oceans, there is vigorous biological research. And throughout the continent, weather data is gathered that feeds climate models.

Indeed, while Antarctica swells in winter and shrinks in summer, its human population does the opposite. In summer they number thousands, mainly scientists and tourists, while in winter only a few hundred hardy souls stay on.

Those who do so are in one of the few places in the world that you can visit without a passport or a visa and where you are unlikely to meet armed force of any kind. Despite the lines on the map showing supposed Norwegian, British, French, New Zealand, Chilean, Argentinian and Australian claims to parts of the continent, only the Latin Americans have shown any interest in a real political land grab. Even their enthusiasm seems to have ebbed as the Antarctic's lack of commercial value has become apparent. Under the Antarctic Treaty, all land claims polewards of 60°S are suspended. In any case, the US rejects them all, which makes them effectively valueless. So these lines on the map will probably continue to have no border guards or customs posts.

The Arctic

At the other end of the world, things are very different. The Antarctic continent is surrounded by the Southern Ocean, the world's wildest ocean, which roars round the Earth untrammelled by land. But while the Antarctic is land surrounded by water, the Arctic is water surrounded by land. The pole itself is merely a spot in the ocean, usually icy but sometimes watery. As we saw in Chapter 6, the Arctic Ocean is probably the most sluggish in the world. It receives the least solar energy to get things going. And it is connected only tenuously to the rest of the world ocean, via the Bering Strait between Alaska and Siberia, and two bigger seaways to either side of Greenland. So it takes in only limited amounts of warmer water from further south.

Unlike the Antarctic, the Arctic has a native human population and the lands that surround it are part of the developed world. They are the northern extremities of Norway, Denmark, Russia, Canada and the US. (Greenland is more or less independent but is in principle part of Denmark.) Some of its peoples, the Inuit or Eskimos, the Lapps and their Siberian allies, have been there for thousands of years and have adapted physically to conditions there (see box). Other northern communities are of more recent origin and are more dependent upon modern technology to stay comfortable and alive.

In contrast to the Antarctic, large areas of the Arctic are in use. Oil and gas are pumped, especially from Alaska and Russia, minerals are mined, and other mostly pretty primary industrial activities go on. In Russia especially, some mining has led to major pollution. In addition, all this activity means that the Arctic contains some substantial towns. Tromsø in Norway has a population of 50,000 and boasts, amongst other attractions, the world's most northerly brewery, university and planetarium.

Life in a cold climate

How do animals and people manage to live in the Arctic? Ask Karl Georg Christian Bergmann (1814–65). This German medic and anatomist worked out that as conditions get colder, animals get bigger.

In the US, a much-studied example of Bergmann's Rule is the **Song Sparrow**. Members of this widespread species that live by the sea are smaller than the ones found in the mountains. The logic is simple. As the animal gets bulkier, it has more cubic centimetres of innards for every square centimetre of surface, so it holds the heat in that much better.

With humans such as the Inuit, Bergmann's Rule applies along with Allen's Rule (Joel Allen, 1838–1921), which states that warm-blooded creatures living in colder environments have **shorter appendages**, again to save heat loss. So the Inuit are stockier and shorter of arm and leg than Africans. They also have other physical adaptations. They have layers of **insulating fat** round the heart and other vital organs. And they have a **faster metabolism** than other races, to keep them warm.

But these adaptations exist alongside an equally important range of artificial measures which are deeply embedded in polar-dweller culture. They have developed homes that stay warm through the long Arctic night. Long before Gore-Tex, and far away from the nearest sheep to provide wool, they developed boots, coats and hoods that keep people dry as well as warm. Living and sleeping en masse, perhaps for months at a time, is second nature. And travellers have noted that just as Africans are expert at finding a bit of shade, so Inuit will find a way of standing in the sun any time it comes out.

Greenland, the frozen island

Of all the Arctic lands, Greenland is the most significant in size and for the world's climate, and most of it lies to the north of the Arctic Circle. (See p.30 if you have forgotten what that is.) It can be regarded as the northern mini-me to the Antarctic. It is only a seventh of the size of the Antarctic continent, with an area of just over 2.1 million square kilometres. But like Antarctica, it is covered in deep ice sheets of immense age.

The Greenland ice cap contains about 2.5 million cubic kilometres of ice. In round figures this means that Antarctica contains 90 percent of the world's ice and Greenland the rest. All the other glaciers in the world are small by comparison.

The ice caps of Greenland are over 4km thick at their maximum depth. Their weight puts so much pressure on the rock below that it has bent downwards into a deep bowl shape, as has happened, too, in much of the Antarctic. In most of Greenland, the rocky basement beneath the ice is below sea level. A column of ice 4km high would put a pressure of over 4000 tonnes on every square metre of rock, because as the snow piles up, the ice lower down is compressed and becomes more dense.

As with the Antarctic ice, this vast amount of material has formed by snow falling and being gradually crushed beneath successive years of snowfall on top. That means that it too can be used for finding out about the past. Drilling has already found ice as old as 200,000 years, from over 3km below the surface. The biggest and best samples are gathered from an evocatively named place called Summit, in the dead centre of the island (see www.geosummit.org). As this is the area from which ice flows outwards, it has the most undisturbed ice, all the way from the surface down over 3km to solid rock. This ice is the prime database for the Earth's climate in geologically recent time, and also tells us about other key environmental variables such as ocean current circulation.

Of course the burning question about the ice caps of both Greenland and the Antarctic is whether they will still be there in a few years, decades or centuries. What are we to make of tales of massive areas of ice breaking off both of these land masses and melting as they float away from the pole?

Part of the answer is that this is a natural process that has always gone on. However, modelling carried out by British environmental scientists suggests that a 3°C rise in average temperatures would melt the Greenland ice cap in time. Some forecasts for global warming contain far bigger possible rises than this. Melting all the ice in Greenland would increase sea

Ice cores

Ice cores are a solid record of the ancient environment. If you want to know how much carbon dioxide there was in the air 50,000 years ago, there is no need to speculate. You can find an ice core that old, find an air bubble in it, and measure its composition directly.

While there are other long-term records of the Earth's environment, such as a 114,000-year run of data from **stalagmites** in Chinese caves, ice cores run back further in time and provide more material to work with.

Ice cores lack some of the data found in the **rock cores** that geologists drill from the solid Earth, on land or in the ocean floor. These contain fossils that provide information about ancient life in a direct way. But ice cores are a uniquely sensitive indicator of the state of the Earth over the last several hundred thousand years.

Ice cores are drilled from areas that rarely or never see a thaw, to ensure that they offer as continuous a record as possible. They also have to come from areas where the ice stays still rather than flowing glacier-fashion. This means that favoured drilling sites are at the centre of Greenland and the Antarctic.

Because the Antarctic gets less snow than Greenland, drilling into the ice there takes scientists back further in time. The oldest ice drilled there, in a 3km core, dates back 740,000 years, while a similar ice core from Greenland would be not much more than 100,000 years old at its base.

Because a 3000m ice core of the sort now being obtained is rare and precious, and weighs about 15 tonnes, it is important to treat it carefully. The US National Science Foundation has a massive cold store for the things, the **National Ice Repository**, in Denver, Colorado (see opposite). Here their properties are measured from the outside and some luckless bits are melted, in the hunt for pollen, volcanic ash, meteorite dust or gas to analyse. Similar studies are carried out in Russia, the UK and elsewhere.

level by perhaps 7m, endangering many of the world's most important cities and a good percentage of its population. As researchers point out, once the ice was gone, Greenland would be far lower-lying and the ice would not re-form any time soon even if global warming were halted or reversed.

...and its relatives

The other major islands of the Arctic are mostly in northern Canada and Russia. Most of them lie between North America and the pole. Their tangled geography is the reason why many European explorers got lost and

Measurements made of such ice cores reveal temperature indirectly, because as it gets warmer or colder, different types of oxygen atom vary in their abundance in the ice. This is called proxy temperature data. Other **proxy data** can be obtained about forest fires and volcanoes, by measuring dust in the ice, and about climate change via carbon dioxide in trapped air.

In recent years, techniques have been developed for finding out about ancient ice without destroying it. One is to measure its dielectric properties, how good (roughly speaking) it is at conducting electricity. Periods when there were a lot of active volcanoes can be spotted in such cores because volcanoes put sulphur in the atmosphere. It ends up as acid in the ice and increases its conductivity.

Although these ice cores are full of data about the remote past, they also tell us about our own behaviour. The zone dating back to the 1960s shows a severe peak in radioactivity because of US, Soviet and UK testing of **nuclear weapons**.

never reappeared from missions to find the **North-West Passage** from Europe to Asia. Many of their names (Devon, Somerset, Victoria, Prince of Wales) reflect the high point of Victorian imperial ambition which was represented by the impractical plan for a sea route from Britain to its Pacific colonies. The area is so remote that substantial new islands went on being discovered in the Canadian Arctic into the 1960s, when aerial and then satellite photography finally produced comprehensive surveys.

The biggest of these islands is **Baffin Island**, now on the adventurous end of the tourist trail as a rock-climbing destination. Like the rest of the Arctic islands, it has been formed by vigorous glaciation and has the peaks, cliffs and fjords to prove it.

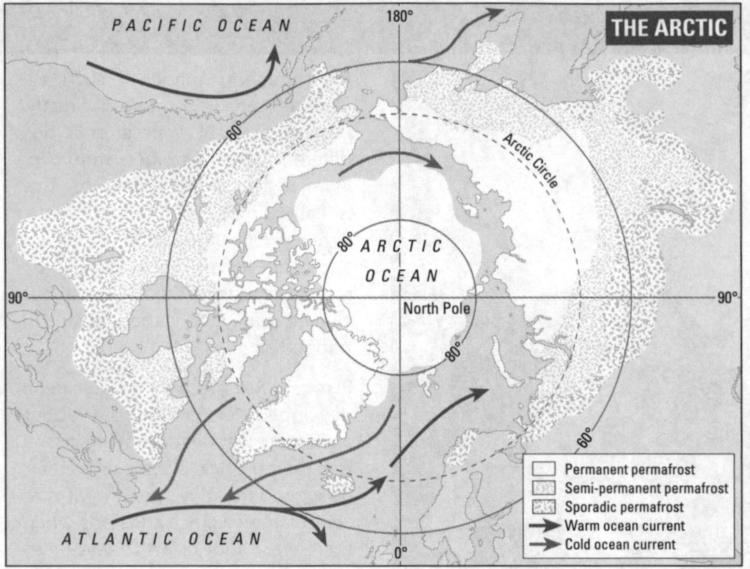

However, the most visited Arctic islands are almost certainly the **Svalbard group**. Politically part of Norway under an elaborate treaty whose signatories include China and Poland, the archipelago has Norwegian and Russian coal mines, but also boasts about 3000 polar bears. So it is fine for country walks for the energetic provided you remember to take a large gun and keep it visible. At a latitude of 80°N, the islands are only 1000km from the pole.

Around the ice: permafrost

The Arctic region's ecology is dominated by the cold. So it is not surprising that even where it is not covered in ice, its land area is often frozen solid for long periods. Any soil that has been frozen for more than two years is known as **permafrost**. That is the definition, but in practice much permafrost has been rock-solid since the last ice age. The miners digging for coal on Svalbard found permafrost hundreds of metres deep and the record is over 1km.

You have probably never seen permafrost, but it occupies about 25 percent of the land surface of the northern hemisphere. About 55 percent of the land in the northern hemisphere freezes for at least part of the year, but only a small part of the land in the southern hemisphere does so because of the dearth of large continents to the far south. As well as much of the Arctic, permafrost covers parts of the Himalayas and patches of the Rocky Mountains.

Scientists point out that despite its name, permafrost is often not very perma. It is often found below a surface layer of soil that melts in summer, making for treacherous conditions for buildings and pipelines.

In recent years, we have come to realize that permafrost ties up about 400 billion tonnes of stored **carbon**. Most of it is in the form of methane, which is three-quarters carbon by weight. These soils have been biologically active in the past, which means that they contain ancient carbon, which is supplemented by fresh supplies from the active layer of soil nearest the surface.

If global warming allowed the permafrost to melt, the released carbon would itself add massively to the greenhouse effect (see p.250). Some forecasts suggest that this will happen in the present century. Human activity produces carbon dioxide emissions of "only" about 25 billion tonnes a year. So the effect of releasing even a small amount of the total in store would be huge. As well as altering the climate globally, melting permafrost would release many cubic kilometres of water now stored as ice. This would alter the northern landscape, such as its river systems, fundamentally. For more on permafrost, see:

CICERO Permafrost page www.cicero.uio.no/fulltext.asp?id=2059&lang=en

Despite all this frozen land, the Arctic is no desert. Its characteristic ecology is **tundra**, a landscape dominated by small plants which succeed because the air is warmer nearer the ground. Further south come milder conditions which support forests, called **taiga** – like tundra, a Russian word.

Tundra (see colour section p.7) is dominated by mosses, lichens and small shrubs, but it supports a much wider ecology including small mammals such as mink and big ones such as wolves. Polar bears, the most characteristic Arctic species, are adept at swimming and fond of fish, but are also happy to eat land animals including humans.

Arctic ecosystems are fragile. The extreme cold has its advantages for science. It means that all those mammoths stay preserved for thousands of years, perhaps with their DNA in such good condition that they can be cloned. But it also means that Arctic environments are especially difficult to conserve. Things happen slowly when the growing season is only about two months long. Damage that would repair itself quickly in a warm climate takes far longer to heal in cold regions. There are now proposals for a sea route from Europe to Asia via the north Russian coast, which would add to the pressure on Arctic environments, especially if it was introduced alongside more oil drilling in the area. New technology has made it possible to travel safely there, while reduced ice cover could make it economically attractive to move big ships on this route for several months a year.

For more on tundra and taiga (and other ecologies) see:

Blue Planet Biomes www.blueplanetbiomes.org

Snowball Earth

Between them, the polar regions make up the majority of what is known to scientists as the **cryosphere**. The term is unnecessary because "the frozen part of the Earth" is just as clear. It is also wrong because the cryosphere is nowhere near spherical. In some parts of the world it is thousands of metres deep, in others completely missing. Once upon a time, however, the term was exactly right.

It seems that 600–700 million years ago, alien astronomers looking this way would have seen "**Snowball Earth**", a world gripped almost entirely by ice. First suggested by Brian Harland of Cambridge University in 1964, Snowball Earth would have looked a lot like Europa, one of the satellites of Jupiter, does today (see p.21), with a surface dominated by pack ice and sea ice even at the Equator. Ice averaging a kilometre deep would have covered the oceans, thicker nearer the poles and thinner nearer the Equator.

In that era, the Sun gave out about 6 percent less energy than it does today, making the Snowball effect more likely. Snowball Earth happened because of "positive feedback" in the climate. If it gets cooler, it snows more. Snow and ice are white. So the more you have of them, the more solar energy the Earth reflects back into space and the cooler the Earth becomes. Things are becoming especially serious when significant areas of the Earth have snow cover that does not disappear over summer but instead remains from one year to the next. This feedback system is called the albedo effect. As we saw in Chapter 1, the albedo of an object tells you how much of the light that arrives there is reflected away and how much is absorbed. Something perfectly white that reflected away all the energy it received would have an albedo of 1, and a perfectly black object that absorbed all the energy that landed on it would have an albedo of 0. The Earth has an albedo of 0.37 today, but Snowball Earth would have been closer to Venus's present-day albedo of 0.65, which makes it the solar system's whitest planet.

One good question is just how the Earth broke out of snowball status once it had got started. The answer seems to be our old friend the greenhouse effect.

Most of the surface processes we see around us such as erosion would cease almost totally if the Earth was frozen solid. But nothing stops volcanoes erupting. Over time the greenhouse gases they emitted, mainly carbon dioxide, would have built up. In fact, carbon dioxide would have accumulated even faster with no plants around to absorb it. As the greenhouse gases built up, the amount of heat that the Earth's atmosphere retained would have increased. Eventually, snow and ice would have melted across significant areas. Then the feedback would have run in the opposite direction and things would have got warmer at a rapid pace. The thick ice may have melted in just a few centuries, and sea temperatures could have reached 50°C. There are signs today of the rocks that would have been laid down from the carbon-dioxide-rich atmosphere that caused the melting. Eventually, however, the ability of the land and oceans to absorb carbon would have increased. As its concentration in the atmosphere fell, temperatures would have come down to something nearer to what we regard as normal.

For more on Snowball Earth see:

Snowball Earth www.eps.harvard.edu/people/faculty/hoffman/snowball_paper

Ice ages

Snowball Earth may be the only time the Earth has completely frozen over. However, there have been periods of more or less severe glaciation throughout the Earth's history. There were major glaciations about 3 billion years ago (the Archaean in geological time), about 2.4 billion years ago (the Proterozoic), and about 300 million years ago (the Permian and Carboniferous), as well as more recently.

Ice signs

Evidence for ice ages comes by the same route as the rest of our knowledge of the Earth – our awareness that the present is the key to the past. Like much else in the frozen world, the story starts in Switzerland. The key character is **Louis Agassiz**, who was born there in 1807 but died in the US in 1873. He was the first person to realize that the valleys of the Alps near his home had been formed by ice during a period when far more of the

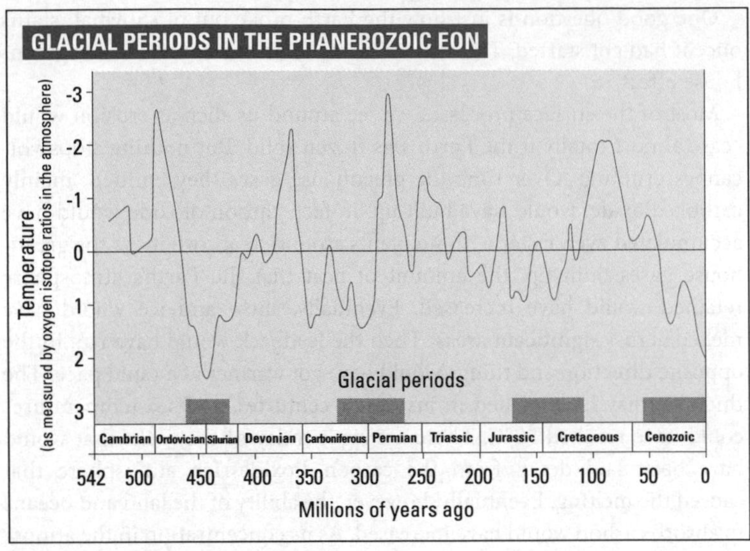

Earth was glaciated than we see today. The evidence he saw is replicated from California to Wales – and indeed equatorial Africa.

What he saw first were distinct **u-shaped valleys**, normally containing a river that is too undersized to have produced it by erosion. Such rivers are known cruelly as misfits and the u-shaped valleys that contain them have been cut by **glaciers**, not rivers. Agassiz (who became a well-known scientist in Europe before being tempted away to the US and becoming a professor at Harvard, starting another trend that still continues) realized that the shape of the glaciers still found at high altitudes in the Alps matched the outline of the valleys thousands of metres lower down. This led him to suggest that they must have formed during an era when glaciation was far more common. He coined the term *Eiszeit* in German to describe it. It is now in universal use in English as **ice age**.

Distinctive u-shaped valleys and size-challenged rivers are only some of the signs of past glaciation which Agassiz recognized. Rocks that have been shaped and polished by a passing glacier are called **roches moutonées**, sheep rocks, because of their general appearance from far off. Look at them close up and something even more interesting becomes apparent. The uphill side is smooth where the ice has passed over it, but the downhill bit is rough where bits have been pulled off it.

Once-glaciated areas also have interesting sights to offer on a bigger scale. Sometimes they exhibit whole rock walls polished to a shine by ice.

Birthplace of a glacier: a cwm at Gryllefjord, Norway

But perhaps the most striking glacial feature is the **cwm**, a Welsh word now in use from Everest to the Andes for the half-cup shape bitten out of a mountain when a glacier forms.

A cwm is a deep mountain bowl formed by the ever-deepening snow and ice that form during an ice age. Over time, the ice turns from a snowfield to a glacier and sets off downhill, gouging out the rock below to form the cwm. Because the glacier tends to get started in an area where snow can survive the summer, they are often on north-facing slopes in the northern hemisphere and south-facing ones in the south, away from direct sunlight. Some such as those on Everest still contain glaciers. Others have lost the glacier that formed there during a previous ice age but still form impressive gouges in big mountains. Their rear slopes are often steep and are a favourite haunt of rock climbers.

Once a glacier gets going it does far more than polish stone. A typical glacier carries rock and other material by the millions of tonnes. Rocks dumped far from where they outcrop naturally are known in the trade as **erratics** (see photo, right). Some are found hundreds of kilometres from home, and can be used to trace the routes which glaciers

Geosight #5: The English Lake District

The Lake District in north-west England has become the beautiful place it is via a complex range of causes. It is on the edge of Europe and has a harsh climate in which its sub-1000m mountains can be as challenging as peaks three times as big in the Alps.

This distinctive area of lakes and mountains comes in two main parts. One, made of softer rocks such as slates, features rounded hills and comparatively easy walking. The other, carved out from igneous rocks, produces craggy landscapes that are more dangerous to the unwary as well as being more spectacular. It includes Scafell Pike, England's highest peak at 978m above sea level.

From about 25,000 to 15,000 years ago, the Lake District was the source of many large glaciers that wended their way far to the south. The high valleys where they began can still be seen today, along with a full array of glacier-related features such as tills, moraines, drumlins and distinctive u-shaped valleys. Where two high cwms from which a glacier began sit alongside one another, a knife-edge ridge such as Striding Edge (see photo, right) is left as a challenge for the walker. Many of the Lake District valleys were cut very deep by ice, and it is these that have formed the sixteen large lakes that give the area its name.

have taken. When a glacier melts, the leftovers are prodigious. Most common are a variety of **moraines** (see colour section p.7). These piles of glacial rock and gravel can be left at the end of a glacier (terminal moraine), at its edge (lateral moraine) or even its centre (medial moraine), when glaciers merge and two lateral moraines meet. Other landscapes distinctive of glaciation are **drumlins** – heaps of material like ancient burial mounds, laid down by ice and then shaped so that their long axis lies parallel to the glacier's line of march. And look out for **eskers**. They are long ridges of material laid down by retreating glaciers and are favoured footways in once-glacial regions.

At the edge of a glaciated area, vast amounts of **gravel** can be laid down, the product of bigger rocks being ground away. These are often quarried heavily for construction. In addition, large areas of once-glaciated land are covered in **boulder clay**. This is a highly distinctive soil made up of clay with large lumps of rock embedded in it and transported long distances by ice. So all in all, a glacial landscape is easy to read and full of interest.

It is more difficult to identify ice ages further back in history. However, amid ancient rocks, it is possible to find **tillites**. A soil such as boulder clay that has been laid down directly by a glacier is called a **till**, and tillites are tills that have been turned over time from soil-like deposits to solid rock. Rocks of the same age also exhibit scrapes from passing glaciers. Geologist Doug McDougall points out that it is by no means simple to identify a full-scale ancient ice age. Several symptoms such as tillites or glacier scrapes must be found in contemporaneous deposits from widely

Loess is more

Glaciers can have a powerful effect across thousands of kilometres as well as nearby. Take **loess**. It is pronounced low-ess, and in small volumes looks suspiciously like dust. In fact its grains are somewhere in size between sand and clay. It was formed in vast amounts by the grinding action of ice during the last, and other, ice ages.

But once formed, loess would not stay still. Instead, its grains were small enough to be blown by the wind, often for thousands of kilometres, until it formed deep deposits in the areas where the wind dropped its load.

Loess deposits now cover significant areas of China, eastern Europe and the US. About 30 percent of the contiguous US has at least one loess layer and often more, and the different layers have been dated to reveal which glaciation episode caused them.

These deposits can be tens of metres deep. The areas where they occur are sought-after because loess is rich in nutrients and forms fertile soils. But there is a down side. Because the material is small-grained and contains lots of shiny quartz crystals it has very little internal friction. In China, the **Loess Plateau** has long been a productive agricultural region, but it has been subject to severe erosion because of the material's poor stability. It is surrounded on three sides by the Hwang He River, whose English name, the **Yellow River**, comes from the vast amounts of loess it carries away. When laid down in thick deposits, it can form unstable steep structures. In China in 1920, an earthquake caused a loess landslide that is blamed for killing 100,000 people, and similar incidents with smaller death tolls continue.

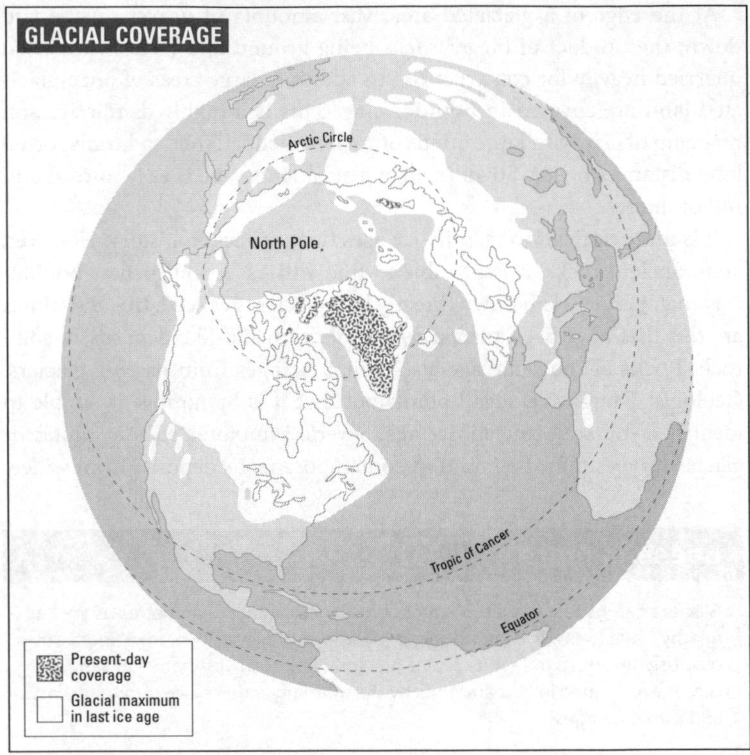

separated parts of the world. It must also be shown that ice was present at low levels as well as in high mountains. A period when ice was only around far above sea level is not an ice age.

Our own icy epoch

Even very ancient ice ages have now been mapped in detail. But most of our knowledge is of the more recent ice ages. The Pleistocene era, which lasted from 1.8 million years ago to about 10,000 years ago, when the Holocene era in which we are living began, was characterized by especially heavy glaciation.

There are a range of theories about why. They include fluctuations in the Earth's orbit and its orientation in space (see pp.32–33) as well as the growth of the Himalayas (see box on p.226) and perhaps increased atmospheric dust caused by volcanoes. Current theory puts changes in the Earth's orbit via the Milankovich cycles in the driving seat.

There seem to have been about 22 major ice ages during this period. The graph below shows the seven most recent, which took place in the last 650,000 years. It plots levels of atmospheric CO_2, which can be used as a surrogate for average temperature. As it shows, these ice ages have a roughly 100,000-year cycle, with most of that period icy in comparison to today and less than 20,000 years warm. Data provided by ice cores from the Antarctic and Greenland for the past few hundred thousand years, and from deep-sea sediment cores from earlier times, means that our knowledge of these climate changes is now very well established. However, exactly which of the low points count as ice ages is to some extent a matter of judgement, especially as some of the low points in more recent times are actually higher than some of the peaks earlier on.

The last ice age ended about 10,000 years ago. The warmer period since then is known as the **Holocene**. All our history and most of our archaeology are to be found in this period. Look out of the window and the Holocene is what you see. Alone among the portions of the geological column, it is still growing. Come back in 100 years and it will be about 1 percent longer.

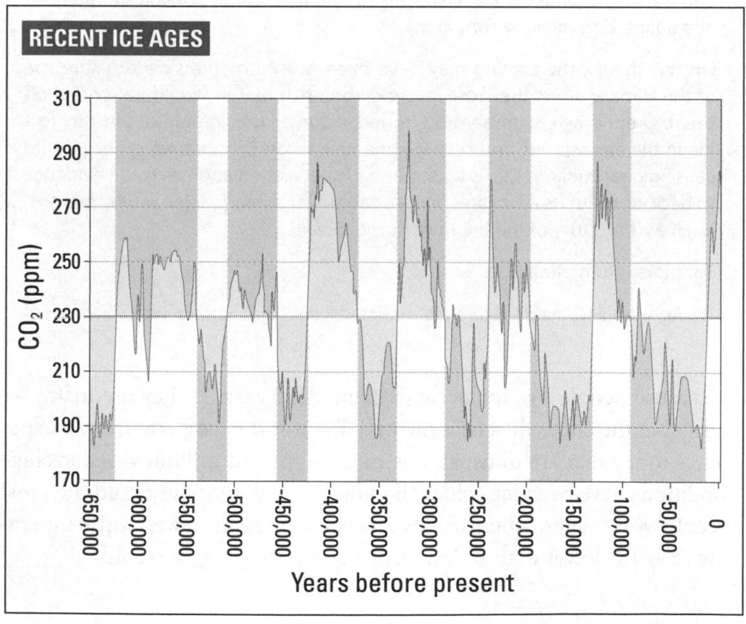

Ice ages and us

What have ice ages got to do with people? Quite a lot, according to current thinking among scientists. Peter Forster, an archaeologist at Cambridge University in England, points out that the successive waves in which humans have spread across the world have mainly been determined by ice ages, both in their timing and in the routes they took.

The route which early humans took from Africa into Asia about 60,000 years ago was probably aided by **low sea levels** (usual in an ice age because so much water has turned to ice) helping them cross into Yemen. The same happened 20,000 years ago when people crossed what is now the Bering Strait between Asia and Alaska to reach America. The end of the last ice age allowed areas of northern Europe from which people had been driven by the cold to be reinhabited.

However, some experts make even bigger claims about the connection between humans and climate. The first accepted stone tools are about 2–5 million years old and come from Africa. Their appearance coincided with climate change in East Africa associated with the spread of plains vegetation and a reduction in the amount of tree cover.

Some scientists claim that the establishment of the present weather patterns in the Indian Ocean, complete with the monsoon that dumps huge amounts of rain in and around India, was accompanied by drier weather elsewhere in the region. Jungles need more water than savannah, so the vegetation changed and walking, tool-using social species of hominids came down from the trees and adapted to the new conditions.

On this theory, the cooling may have been caused by the slow but sure rise of the Himalayas. As they rose, more carbon-rich matter would be eroded off them, swept down to the sea and buried, reducing the amount of carbon dioxide in the atmosphere and setting up a reverse greenhouse effect that would cause global cooling. This is a neat line of speculation and has some evidence to back it up. But its advocates are probably overplaying their hand by presenting it as "the" story of the origin of *Homo sapiens*.

For more information, see:

Ice Ages and Human Dispersals www.mcdonald.cam.ac.uk/genetics/iceage.pdf

Because people live for perhaps a hundred years if they are lucky, we think that the era we live in is normal. But it is the temperatures we experience today that are unusual. For most of the last million years, average conditions have been far colder than our idea of normality. And the most recent two ice ages, when sea level was about 120m lower and temperatures 5–10°C lower than today, were also the deepest on record.

However, as the graph shows, the current warm spell is nothing special. And on past form, you might think the line is due to head south again relatively soon. That is why geologists call it an **interglacial** – a period between ice ages – and speculate that we may be nearer in time to the next ice age than to the last one.

Even if a full-blown ice age does not happen soon, the temperature might not remain at its current balmy heights. During the most recent ice age there were two cold snaps, beginning about 14,000 and 12,800 years ago, now called the Older and Younger Dryas. They are named after the Dryas flower – still found today in tundra ecology, but far more widespread during these two periods. The Younger Dryas, about which we have more detailed information, lasted about 1200 years, and began and ended very suddenly. Seen alongside everything we know about the Little Ice Age, a cooler spell during the current warm period (see pp.34–35), it suggests that sudden, brief cold snaps may be more common than had been thought.

Melting glaciers

As recently as 1981 the maverick astronomer **Sir Fred Hoyle** published a book called *Ice: The Ultimate Human Catastrophe* predicting a new ice age. One possibility he considered was that an ice age might be triggered by a large meteorite impact putting enough dust into the Earth's atmosphere to cause significant cooling. At the time Hoyle wrote *Ice*, there was also an increasing awareness that the use of nuclear weapons could have the same effect: they might start fires that would cool the Earth by putting large amounts of smoke and dust into the atmosphere, a scenario called **nuclear winter**. Hoyle added that a giant volcano or an asteroid impact could have the same effect. He thought this might happen within a decade. However, only a couple of decades on from the book's publication the current fear is quite the opposite – that the Earth may be heating up. What does this mean for the icy regions of the Earth?

Tales of retreating glaciers have been increasingly in the news in recent years, either because glaciers are retreating more or because the media have caught on to the idea – or both, this author's favoured explanation. The term that the glaciologists like to use is "**mass balance**", the gain or loss of ice from a glacier. The story is by no means simple. The glacier cover of the Alps may have diminished by up to 40 percent in area since 1850, or 50 percent in terms of mass. But in Scandinavia – mainly Norway

White gold

Despite their decline, it is worth celebrating the great glaciers of the Alps instead of mourning them. There are glaciers all along the arc of the Alps from France to Slovenia, including the **Mer de Glace** in France, the **Stubai** in Austria and dozens of smaller ones. They provide the headwaters of many big European rivers, but in addition the snow and ice of the Alps means leisure for millions of people. The mountains are perfectly placed at the centre of Europe, a short journey from many major cities. Despite many areas of the Alps being designated as national parks, the amount of use they get is almost as threatening to them as climate change.

The number of people who want to visit the Alps means that the snow and ice is worth money. Oil may be black gold to Texans, but Austrians call snow **white gold** with equally good reason. In one slightly astounding manoeuvre, the Austrians are experimenting with a range of materials that can be used as overcoats for the Stubai Glacier, to keep it fit for skiing. The idea is to spread a thin layer of very white material on the snow when the Sun is at its hottest, to reflect away the most intense heat. It may preserve both the snow and the incomes of the ingenious locals, but is not a general solution to the problem.

– there seems to be more ice rather than less in the biggest glaciers. However, the total equation since 1960 seems to suggest a mass balance 6000 cubic kilometres in debit across the glaciers for which there are sensible data, including those in most of the major mountain regions of the northern hemisphere. It is not likely that these are a statistically unrepresentative sample. For more on glacier mass balance see:

SOTC Glaciers page http://nsidc.org/sotc/glacier_balance

Some of this change is probably natural. As we have seen, we are living in an **interglacial**. The glaciers will probably go on retreating until they start to grow again as the next ice age sets in. But there are estimates that another 16cm of sea level rise could be caused by glacier melting during this century.

Glacier surges

The great glaciers of Alaska and northern Canada have been observed displaying "**surges**" in which they advance by many tens of metres a year. But this is not a sign of expansion. It often happens because they are being undercut by streams formed as they melt, which lubricate their progress and hasten yet more melting. An example is the **Bering Glacier**, the biggest in North America. It surged and then retreated several times during the twentieth century and lost 130 of its 5200 square kilometres of area. Between 1993 and 1995 the glacier moved forward by about 5km and satellite positioning has been used to measure the thinning of the ice further up the glacier as it moves downwards or melts. Seafarers might rejoice at losing this and other "maritime" glaciers, those that end in the sea. They are the prime source of **icebergs** in the oceans further south. But the local environmental effects are far less welcome.

There appear to be several ways in which glaciers surge. Tavi Murray of Leeds University in the UK says that some surges are long and slow, lasting over a decade and occurring at a pace of "only" 2–5m a day. They happen when water saturates the soil beneath a glacier and its grip on the ground below is weakened. The other type of surge lasts a year or two but can involve movement of up to 50m a day. This happens when the normal water flow beneath a glacier is diverted into cavities in the ice and large amounts of pressure build up.

Roof of the world

People who worry about climate change tend to talk about what would happen if the Antarctic ice sheet or the glaciers of Greenland melt. And as we have seen, these ice masses are so huge that their melting would mean major environmental damage via sea level rise and other effects. By contrast, the rest of the world's glaciers contain less water and their disappearance would be a regional rather than a global disaster.

If the Earth is losing its glaciers, the place where it matters most is in the **Himalayas**. The Himalayas are a mountain chain 2500km long, with innumerable subsidiary chains such as the Hindu Kush. They contain the world's highest mountains including all fourteen that are more than 8000m high. The area contains an estimated 67 peaks higher than Aconcagua in the Andes, at 6960m the highest mountain not in the Himalayan region.

As well as being a stupendous mountain system in their own right, the Himalayas are backed by the **Tibetan plateau**, one of the world's biggest upland regions. Most of it lies about 4000m above sea level. A new Chinese state atlas of the country's icy, desert and tundra regions claimed in 2005 that China had nearly 60,000 square kilometres of glacier. Most of this is in the politically disputed area of Tibet – or "the country's west", as the state news agency euphemistically puts it.

Their height and size mean that the Himalayas help shape the Earth's climate, as we saw in Chapter 5 (see pp.153–54). And just as the Andes and the mountains of East Africa give the world the Amazon and the Nile, the Himalayas are the source of some of the world's great rivers, including the Indus, the Ganges, the Yangtze, the Mekong and the Brahmaputra.

Immense glaciers surround all the major Himalayan peaks. As with other glaciers in the world, they are a navigational hazard to climbers because the ice is not solid and forms crevasses that can trap the unwary. Crossing them is a major part of Himalayan climbing, as the many accounts of crossing the Icefall, the glacier on the main route up Everest, suggest.

It is not likely that this much ice is going to vanish in the next few decades. But there is every chance that these glaciers will go on shrinking at a far faster pace than they have in recent centuries.

How to start an ice age

The Earth's 160,000 glaciers may be culled radically during our lives, but it is still very likely that the Earth's long-term future contains more ice ages. If it does, the massive glaciers and ice sheets that it entails will start small, like those we see today once did, with snow crystals that fall and do not melt.

As we saw in Chapter 5, **snowflakes** form in clouds, and far more form than ever reach the ground. Often they melt within the cloud or on the way down and fall as rain.

But the one thing everyone knows about snowflakes is wrong. It is not true that every one is different. In fact they come in a few basic shapes and sizes. Some considerable effort has been put into classifying the different types – in 1951 the International Commission on Snow and Ice produced a snowflake classification, and more recently an eighty-type classification was introduced. The bare bones are that the simplest shapes are hexagonal **prisms**. Depending on their proportions, these can appear as long columns (like a pencil) or thin plates. Then come the "**stellar**" plates, which have six arms forming a star-like shape. If these arms develop branches, the crystals are known as **dendrites** (tree-like shapes). The bushiest and most elaborate of these are called **fernlike**.

The shape of a snow crystal depends on both the temperature and the humidity of the air in which it formed. Even small changes in **temperature** can make a surprising difference to the shape of crystal formed. So thin plates and stars form at around -2°C, while at around -5°C, columns and slender needles form. When the temperature drops to around -15°C, plates and stars form again, while at temperatures down to around -30°C, plates and columns are found. Why the shapes change so much with temperature remains something of a mystery. Kenneth Libbrecht, the world's leading expert on snowflake sci-

Scientists funded by the British government have looked at what this might mean for the Himalayas and their neighbours. It turns out that the **Indus**, the principal river of Pakistan, could carry anything from 14 to 90 percent more water in the next few decades. This would mean massively increased flooding and erosion. But once a significant percentage of the glaciers has melted, the amount of water in the river would fall by 30–90 percent over the coming century. As this water is the basis of all life and agriculture in a fast-growing nation, the effects scarcely need spelling out.

Pakistani politicians have used these predictions to back plans to capture more water behind the big dams that environmentalists hate. Other rivers such as the Ganges could also be carrying 50 percent less water.

ence and chairman of the physics department at the California Institute of Technology, no less, says that nobody knows just why.

When **humidity** is low, the simplest plates and prisms are formed, while the more complex or extreme shapes, such as the elaborately branched dendrites and very slender needles, form when humidity is high.

For more information and images, see:

SnowCrystals.com www.its.caltech.edu/~atomic/snowcrystals

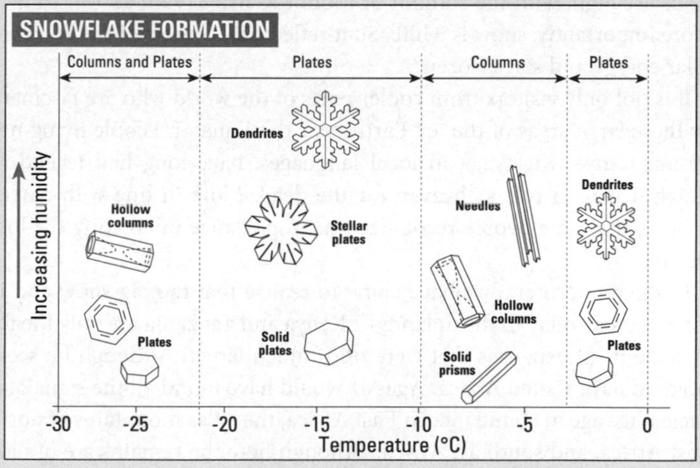

These changes would reflect big alterations in the environment at high altitude. Small glaciers would vanish and big ones would retreat. There are signs that this is already happening. In the Garhwal Himalaya, the Gangotri Glacier has been retreating since 1780, but the process has sped up since 1971. It has lost 850m of its original 31km length since then.

On the Tibetan plateau, the Meili Glacier has retreated by about 20–30m a year since 1998. In 2003, scientists calculated that the average temperature in the area had risen by only about 1°C over thirty years, not much by the standards of global warming. This may suggest that other forces are at work. But they add that there is a possible mechanism for the change. If water that once fell as snow now lands as rain, it will erode the ice instead of adding to it. The same story applies to many other of the 46,298 glaciers said to be on the plateau. According to Japan's *Asahi* newspaper, the volume of water coming off them has increased by 20 percent over the last 40 years and their area has dwindled by 7 percent.

Snow at the Equator?

Glacier retreat is also going on in more unexpected parts of the world. Nobody believed the eighteenth-century European travellers who first reported snow at the Equator in Africa. They may have been wrong about the dragons and the unicorns, but on this at least they were right. **Mount Kenya** is on the Equator and has permanent snow.

The snow stays there for a number of reasons. One is that the snowfields are high, with the summit of Mount Kenya 5199m above sea level. More importantly, snow is white. So it reflects away most of the incoming solar energy and stays frozen.

It is not only visitors from cooler parts of the world who are fascinated by these large areas of the icy Earth near the Equator. People living near Mount Kenya, Kirinyaga in local languages, have long had legends in which it plays a role as heaven for the dead. More in line with current science, the same people recognized its importance in creating the local weather.

In the modern era, we have come to realize that the big snow and ice zones we see today in the uplands of Kenya and Tanzania are only the fossil remains of expanses that were once much larger. Although he seems never to have visited Africa, Agassiz would have found all the signs of an ancient ice age in abundance in East Africa, the Atlas mountains of northwest Africa, and southern Africa, although here the remains are of older ice ages. Nor is snow at the Equator confined to Africa. In 1632, the Dutch sea captain Jan Carstensz sighted more of the stuff just south of the line on **Irian Jaya**, the easternmost part of what is now Indonesia. The peak he saw is still called the Carstensz Pyramid, 4884m high. And of course, the Equator also passes through the Andes, the planet's most prominent mountain range after the Himalayas.

Although the persistence of ice in these areas is striking, it should not be taken for granted. Scientists looking at Mount Kenya report that the mountain and its surroundings have a dozen glaciers left, but they are all retreating fast and others have already gone. The mountain was once well over 6000m high, before severe glacial erosion took over 1000m off it. Some time soon, the glaciers that did this could be gone. **Kilimanjaro** in Tanzania, Africa's highest peak at 5895m, is reported to have lost 80 percent of its glaciers in the last century. The same story could be told from Norway to Argentina.

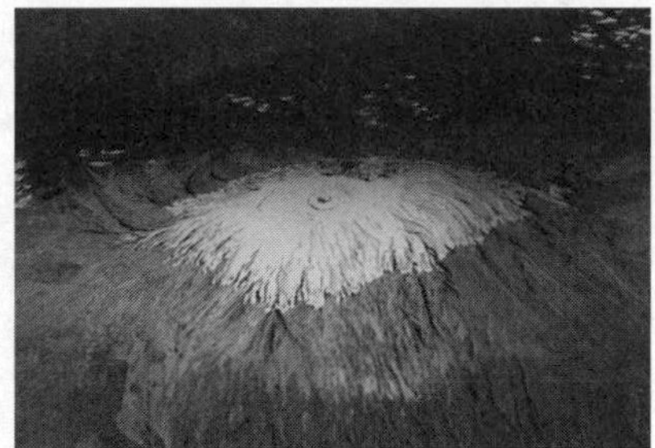

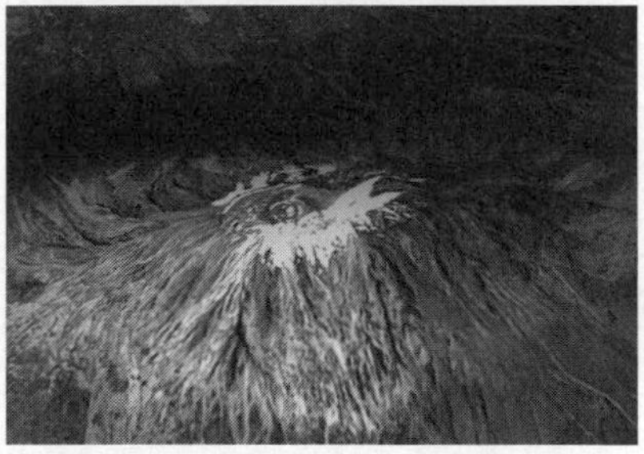

Snow at the Equator, but for how long? Kilimanjaro, Africa's highest peak, in 1993 (top) and 2000, showing extremely rapid loss of snow and ice cover.

If ice fields do vanish, other things change too. Many of these environments have Alpine-related ecology containing species that are unknown elsewhere. Perhaps more importantly, the **Nile**, the world's longest and possibly most important river, is fed by lakes that themselves are filled by water from the great mountains of East Africa. As we saw (see p.200), the Nile's water is already in too much demand for its supply to be endangered.

8

The Earth and us

8 The Earth and us

Many a literary metaphor contrasts the unvarying Earth with the fast-flowing fortunes of people. If you've read this far in the book, you'll know that this is inaccurate. Weather systems, sea levels, ice sheets and river courses change over time, in some cases surprisingly quickly. Of course, the solid Earth is made of sterner stuff and changes more slowly. But eventually, mountains are worn away, oceans open and close and continents are swallowed up. In some parts of the world, earthquakes and volcanoes can remake landscapes in days.

What is true is that the Earth is a lot less unvarying than it used to be. In recent times, a new and vigorous Earth-altering force has emerged that transforms its surroundings over years rather than millennia.

Yes, I am talking about you.

People have always changed their environment. But in recent centuries, things have become much more serious. Part of the issue is a simple question of **population**. A couple of thousand years ago, Jesus was one of about 300 million people at most. In 1800 there were 900 million humans, and in 1900, 1.6 billion. Now there are over 6 billion of us, compared to only about 200,000 of our nearest relative, the chimpanzee *Pan troglodytes*.

The other reason for our increasing domination is **technology** and the speed it adds to human action. We are able to use more land, water, minerals and other resources than we could even a few centuries ago. It takes a forest decades to spread a few kilometres, but people need only a few days to cut it down.

This combination of a burgeoning population and increasing impact per person means that the Earth is turning into a unique case, as far as we know, of a planet that is consciously managed by a single self-aware species.

Population growth

Is it a rule of nature that the population keeps on increasing? It might seem so from the relentless graphs of world population that adorn most books on the subject, including this one.

But this century's newspapers seem to contain more headlines about a dearth of people than about a surplus. The reason is that people in the developed world are living longer and are seeing less reason to replace themselves. In 2006, there were four times as many people of working age in the developed world as there were over-65s, and it is predicted that the ratio could reach three to one by 2020. The most extreme case is Italy, where there could be less than two people of working age for each pensioner by then.

The moral panic over people living longer is rooted in the financial fears of pensions providers and health services, which worry how they will cope in this new world.

This is not *The Rough Guide To Pensions And Health Provision*, so we do not need to solve that problem here. But the fear it encapsulates does contain an important message. Take the US as an example. Every year about 8 people die out of every 1000 in its population, while 14 are born. Result – population growth of 6 people per 1000. So far, it has turned out to be easier to make people live longer than to get them to reproduce less.

But it seems that the population of the Earth reached its maximum rate of growth in about 1990. Since then, changes such as the improved status of women in many societies and the availability of contraception have reduced growth. If these trends continue, world population will grow to about 9 billion during the present century but will be on the way down by the century's end.

This forecast contains a huge number of assumptions. One is that there is no major disruption to social progress, such as an asteroid strike or nuclear war,

It is best not to get carried away by this idea. Despite the sway humanity now has over large areas of the Earth, it would be wrong to think that we have it under any sort of control. The increasing death toll from **natural disasters** makes it clear that we are far from being in control of the inanimate world. And as the constant flow of new infectious **diseases** shows, from HIV and Ebola to each winter's new flu, nature is still tougher than we are and is producing new forms of life the whole time. Indeed, our new technologies have actually aided the global proliferation of diseases. It is now possible to get from one side of the world to another in a day, which is less than the incubation time of many diseases, and new forms of influenza and other diseases make the most of this, getting from Asia to Europe or the US with little warning. If you live near a European air-that would cancel all bets. More subtly, it assumes that general Western-style social progress will be embraced in areas such as Africa which have missed out on past decades of development.

If population peaks at about 9 billion, humans will still be by far the most numerous species of large land animal on the Earth. The issue is what sort of demands they will be making on the rest of the Earth. By that time, it is likely that the high days of oil and gas consumption will have passed, though perhaps not of coal. It is certain that most of the people alive will be living in cities. It is very possible that they will be living in a world whose climate is tangibly warmer than today's. Whether we are clever enough to build a sustainable world on this foundation is less clear.

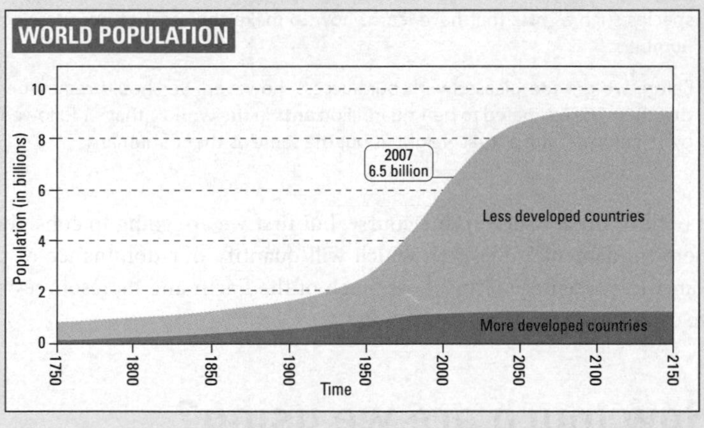

port with frequent flights to Africa, you can get "airport malaria" from mosquitoes that have arrived by plane. More serious plant and animal diseases take advantage of the increased world trade in meat, vegetables, pets and livestock to spread around the globe. Even in the city, humans are not in total charge of their environment and share it with other species, something I am reminded of every few days in inner London when I see a fox in the garden.

However, although we may not always have the upper hand, we're certainly making our mark on the world around us. Much of the attention that is paid to our impact on the Earth is currently focused on the question of human-induced **climate change**. Another issue of serious concern is the number of species that we are driving to **extinction**. We will look

Sheer weight of numbers

Try thinking quantitatively about the hold humans have on the world. There are about 400,000 African and 40,000 Asian **elephants**. At about seven tonnes apiece, this means that the world has 3–4 million tonnes of elephant. This is about 1 percent of the mass of humans.

Throughout history, human beings have been reluctant to share their environment with any other large undomesticated animal. Maybe they are dangerous, perhaps they compete with us for space and food, or maybe they just disturb our idea of our own importance. The best place to find them now is in the oceans – hence the popularity of **whale** watching – and in game reserves in Africa. There is certainly no large wild land animal that exists in numbers remotely comparable to the billions of humans. The most successful mammals, judged by numbers, are the **farm animals** that humans keep, or species such as **rats** that have learnt how to make the most of our growing numbers.

Before we get too pleased with our success, however, we should remember that there are estimated to be 1 quadrillion **ants** in the world – that's 1 followed by 15 zeroes – with a total weight about the same as that of humans.

at both of these issues in due course, but first we are going to consider a more fundamental question, which will quantify our dominance of the planet in percentage terms: how much of the Earth and its resources are we exploiting for our own purposes?

How much are we using?

Let's start with the basic process that drives life on Earth – **photosynthesis** (see box). Just how much of this process humans are making use of is controversial. As long ago as 1993, one estimate put the proportion of land plant production being used by humans at 40 percent. The jargon term here is NPP – **Net Primary Production**. The percentage of NPP which humans are using must have grown since then, as jungles have been cleared for farming in Brazil, Africa and elsewhere. But despite the rise in world fish catches, the share of marine biological productivity that people are using is far lower. More recent calculations by the University of Michigan suggest that humans are using (or have killed by urbanization and other development) between 31 and 39 percent of the biological productivity of the land area of the Earth. But as we are only using about 2 percent of the productivity of the oceans, we are "only" using about 25 percent of the planet's total biological production. As environmentalists point out, this is a lot for one species out of around 30 million that exist on the Earth.

Give me land, lots of land

Another way of looking at our exploitation of the Earth's resources is to think about the amount of the Earth's **land surface** that we are making use of.

Humans have been altering the landscape to suit their needs throughout history, and it is sloppy thinking to believe that only modern Western technology can have a big effect on the terrestrial environment. In most of Britain, for example, political stability, population growth and rising wealth led to large areas of the landscape being captured for food production in the seventeenth and eighteenth centuries. Now only very limited parts of the highlands of Scotland are genuinely wild, in the sense that they have not been restructured by human use. Areas such as the English Lake District are precious and beautiful, but if it were not for people and

Photosynthesis

Photosynthesis is the difference between the Earth and everywhere else we know about. From Mercury to Pluto and, so far as we know, on all the hundreds of other planets we have now discovered, sunlight – or starlight – falls on craters and mountains, clouds or ice, and is either absorbed or reflected. But on the Earth, it is absorbed by purposeful systems that do something with it. That something, photosynthesis, happens on a molecular scale but has planetary effects.

Photosynthesis is the process whereby light is absorbed by plants, algae and other organisms and its energy is stored by a chemical called adenosine triphosphate, ATP. In plants, the solar energy is absorbed by a pigment called **chlorophyll**. It absorbs most of the light apart from the green. The green light is reflected, which is what makes plants look green. Other organisms that perform photosynthesis, especially some bacteria, use other pigments to absorb the light and can be red or blue.

The light absorption is the "light phase" of photosynthesis. It is followed by the "dark phase" in which energy stored in ATP is used to turn **water** which plants draw up from their roots, and **carbon dioxide** from the atmosphere, into **sugars**. This reaction creates living matter and also releases **oxygen** for animals to breathe, including us. So it is the vital transaction in the existence of life on Earth.

their sheep, they would look very different. Look at the walls you see running along 1000m or more above sea level in the Lake District or North Wales. They are there because sheep have been put on the hills in numbers, removing some plant species and encouraging others. In Scotland, sheep were the main reason for the removal of large areas of forest and much of the human population during the ruthless **Highland Clearances** of the nineteenth century. Much of the rest of Europe – aside from such areas as the peaks of the Alps and the glaciers of Norway – has been similarly altered by human hand. And even less technological societies can alter the land fundamentally, as Australia's Aborigines did by widespread burning of vegetation.

Jim Williams of the University of California in Berkeley points out that the Earth's land area is about 150 million square kilometres. Of these, about 50 million are not good for much – Antarctica, the Sahara and the like. Another 50 million is **forest** and another 50 million is other potentially productive land. Of the forests, 5 million square kilometres are "managed" for human use. Of the other 50 million square kilometres of usable land, 25 million are in use for **agriculture**, of which 15 million are used for crops and 10 million for grazing. Although it might not feel that way, the area taken up by **cities**, **dams**, **mines** and the like is comparatively small, under 2.5 million square kilometres. Put together into a perfect square, it would make a world city about 1600km on a side.

Williams estimates that about 15 million square kilometres of natural woodland and grassland have been converted to human use, mainly since 1700, while a similar amount has been turned to desert, a process that continues today at a rate of about 60,000 square kilometres a year. Forest destruction is continuing at a rate of 150,000 square kilometres a year in the Amazon basin alone, with more elsewhere.

Does this matter? If land that used to be an unmanaged forest turns into a sugar plantation, is the Earth any worse off? James Lovelock of Gaia fame (see box on pp.256–57) points out that most of the areas of previously wild land now being converted to human use are near the Equator. As he says, areas nearer the poles have far fewer species of plants and animals, because every time an ice age comes along, they are bulldozed by glaciers. The warmer parts of the world escape this depletion and have the task of repopulating the polar regions as the ice retreats. Even *Homo sapiens* joined in after the most recent ice age by repopulating Europe.

This means that converting equatorial land to agriculture has a high cost in **species wiped out**. It also damages some of the systems that do most to keep the Earth stable. As we have seen, plants are based on photo-

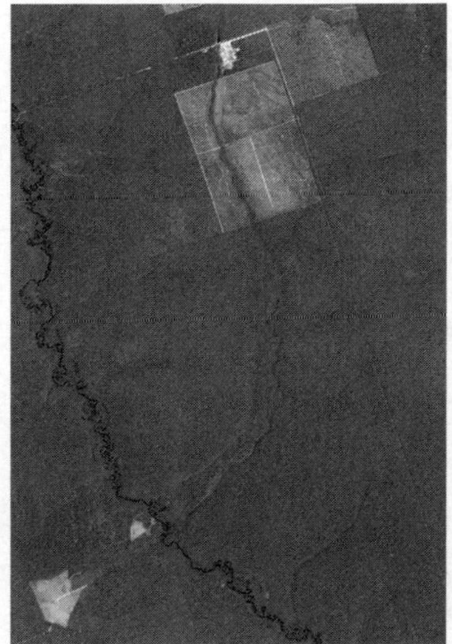
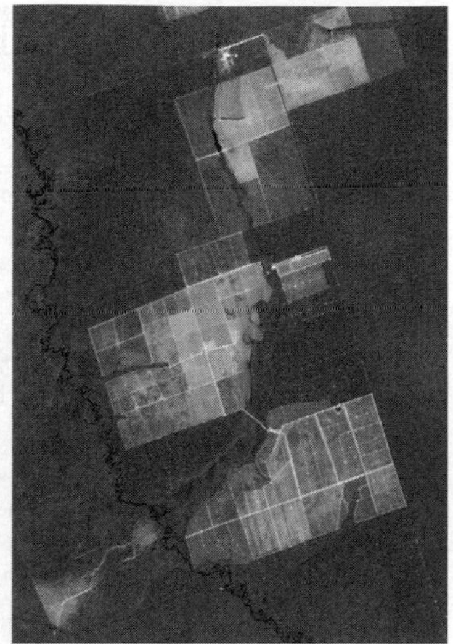

Running out fast: forest being cleared and made into fields in Mato Grosso, Brazil, between 2001 and 2006

synthesis. The tropics are as green as they are because they have plenty of sunlight and water to drive the process. So they provide very lush and diverse environments. But many **failed agricultural projects** in these areas suggest that it is not safe to assume that cultivation can take over this land successfully. Often severe soil erosion and other unexpected effects result in ecological damage and economic failure.

How much are we eating?

The University of Michigan scientists reckon that we are eating about 800 million tonnes of plants a year and feeding a further 2.2 billion tonnes to the animals we eat. This means that if you believe that the world is going to run out of food, a mass switch to **vegetarianism** is the easiest way to avert the problem. But bear in mind that a lot of the land used to graze animals, especially sheep, is not much use for crops. Only about 10 percent of the Earth's surface is **arable** – usable for ploughing, from the Latin word for plough, *ara* – although irrigation, the development of new crop species (perhaps through genetic engineering – see box on p.244) and other forms of technological advance can push this percentage upwards. We are getting through 2.2 billion tonnes of wood a year, for firewood and building, and 2 billion tonnes of fish.

Genetically engineered crops

As the pressure on our food supplies increases, could genetically engineered crops help food production keep pace with demand?

The term genetic engineering, or **biotechnology**, normally means taking genes from one species and putting them into another. Almost all the plants and animals we eat have been improved by gradual breeding processes. Biotechnology allows the time needed to make such changes to be cut. And it has allowed new properties, such as herbicide tolerance and pest resistance, that do not exist in a species to be built in. Climate change could mean that **salt and drought resistance** might be valuable additions to many crops in the future.

Genetically modified crops, mainly soya and maize, are already being grown around the world, the majority of them in the USA. American consumers have been eating them without much complaint for a decade or so now. However, European consumers reject the idea of GM, and make trouble for politicians and companies that try to tell them GM is good for them.

They may have a point: if GM does catch on in a big way, it is certain that the unintended effects will outweigh the intended ones. It is not in plants' nature to stay where they are put, and GM plants are sure to compete with natural species in the wild. In addition, thanks to the work of Lynn Margulis and other scientists, we now appreciate that species swap genes the whole time via micro-organisms. So a gene that biotechnologists have installed in an apple tree to help fight pests will get into a tree in the wild and make it pest-proof too.

This means that people are using 7.2 billion tonnes a year of biological products from the land and oceans. This is just over a tonne per person, which sounds like a lot, but is only 3.2 percent of the 225 billion tonnes of plant matter a year that the Earth produces. However, agriculture is not 100 percent efficient. Add to these totals the losses inherent in growing all those crops and trees, and it turns out that we are getting through over 40 billion tonnes of biological production a year. This figure is bound to rise drastically if the billions of people in India and China start to consume something like the amount of food, wood and other biological products that people in Europe, Japan and North America take for granted.

In 2002, we ate our way through 242 million tonnes of **meat**. To put this in perspective, if the 6.5 billion people on the Earth weigh perhaps 60kg each on average (remember that many of them are children) that means that there are 390 million tonnes of human being. Eating as much meat as we can is one of the first things that happen as people get richer. According to the admirable Worldwatch Institute in Washington DC, meat consumption doubled between 1977 and 2002. As countries such as China and India get richer, this figure is bound to rise much further. Indeed, Chinese people are already developing an increasing appetite for beef. All those cattle need good grazing land, which has to be well watered, but because beef can command a high price, this use of land is bound to outbid products of traditional, less resource-intensive, agriculture.

Drinks all round!

In 2000, 2800 cubic kilometres of water were used for **agriculture**, 800 for **industry** and about 600 for **domestic use**. This last figure might miss some water gathered and used locally in the developing world, but the total under this heading cannot be very great. The total, 4200 cubic kilometres, might sound like a lot, but it is only a tiny percentage of the fresh water on the planet.

So why the constant news stories about water shortages? The problem: geography. With a world-traded resource such as oil or coal, it might be worth thinking how much of the available material we are using per year. But there are no sharp-suited water traders making fortunes by speculation in London or Chicago. We have to think about the issue in regional rather than global terms. Thus by one count, Canada has 20 percent of the Earth's fresh water and 0.5 percent of the people. China has 7 percent of the water and 21 percent of the people. Unless most of them move to Canada there will be growing stress on rivers and groundwater in China. The same goes for Europe, large parts of North America, and growing areas of Asia.

In some areas, old **groundwater reserves** are being mined far faster than they can be replaced. Populous, dry areas such as the Middle East are especially prone to this unsustainable behaviour (see p.198). Some of the aquifers that are being depleted formed in the past when the climate was wetter in these areas, and are essentially fossils. But the special paper on groundwater produced for the planned International Year of Planet Earth (set to be in 2008) points out that some of these reservoirs are very large. One, the Nubian Sandstone Aquifer, underlies parts of Chad, Egypt, Libya and Sudan and contains about a century of humanity's current water use.

Many observers have pointed out that water shortages in the developing world have severe effects on health and economic growth. Solving these problems will mean more water abstraction. But there is also scope for improved practices. By comparison with most economic activity, water use is exceptionally **inefficient** in most countries. Economists say that this happens because water is regarded as cheap or free in many parts of the

Want to save the planet? Move to the city

Although it is natural to regard cities as the most unnatural and environmentally damaging places on Earth, perhaps there are two sides to this story.

Throughout history, people have voted with their feet by moving to cities. They have often faced extreme exploitation and poor living conditions to get closer to the jobs, education, crowds and culture that cities offer. In the recent past, **urbanization** has been gaining strength across the world. In 1950, 30 percent of the world's population lived in cities, while in 2006 the total reached 50 percent. That adds up to 3.3 billion people.

Indeed, documents produced for the 2008 International Year of Planet Earth point out that it is not just cities that are growing, but "**megacities**" of 5 million or more people. There could be sixty of these by 2015, and they would be home to over 600 million people. Many big cities dominate their countries. In the United Kingdom (according to the 2001 census), over 7 million people were living in London and fewer than a million in Birmingham, the second city. Even that figure understates the size of London by only counting people within its boundaries as they existed some decades ago.

There are problems with urbanization and the main one is that it never knows where to stop. Many cities are so popular that they just keep growing. They tend to expand, as outer London has done in the Thames Valley, onto the **flood plains** of rivers. Elsewhere, **avalanche-prone areas** or regions with endemic **subsidence** get built upon, usually by the poorest inhabitants.

And although we know a lot about the social and economic advantages of cities, our knowledge of how to make them more physically friendly for their inhabitants is still underdeveloped. As the Year of Planet Earth people tactfully put it, "In the developing world, megacities grow faster than their infrastructure." They might have added that the rich world has its share of this problem too, as anyone who has tried the subway systems of Tokyo or New York at 8am knows.

Since most people want to live in one, lessening the environmental impact of cities is a top priority for reducing the effect we are having on the Earth. However, in principle, there are good reasons why urbanization might actually be beneficial for the planet. People in big cities ought to have **shorter journeys** to work, fun or the shops. They can be connected to electricity and fuel supplies by shorter power lines and pipes. They can get **economies of scale** in the raw materials they use and in the technology used to recycle or remove their leftovers. And they can use **public transport**. Country-dwellers in the rich world have to start every activity by getting into a car.

Of course, the fact that most people live in cities means that most people do not live in the country. The growth of both megacities and more modest urban centres means that despite population growth, there could be fewer people, not more, in **rural areas** in future. This itself might be a good thing for the planet.

As we have seen, most of the Earth is pretty empty. Only 30 percent of it is land, and of this, only about 10 percent is used for agriculture and another 8 percent for cattle rangeland. Take a look at Japan, which has 127 million people. Most of them live along the southern coastal strip, especially in the Tokyo megacity. Much of the country is empty, particularly the mountain chain that makes up most of the centre of Honchu, the main island.

The world already has substantial **national parks** and reserves. Even in the crowded UK, they add up to over 10 percent of total land area. In the past their protection has been patchy. One – Snowdonia in North Wales – contains a nuclear power station. One way of keeping the Earth in good shape would be to increase the area of it given over to such parks and reserves, in which people are scarce and go only to take a look, not to carry out primary economic activity unless it has been pre-approved by centuries of proven low-impact success.

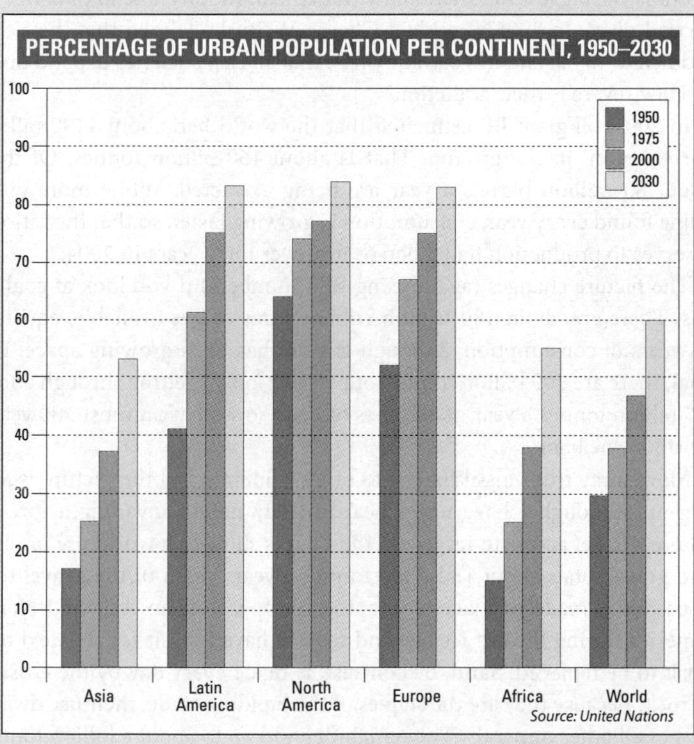

world. Whatever the reason, the UN says that about 60 percent of irrigation water gets lost before it sees a field, and that in some cities, a third of the water gathered escapes before it gets to the tap. Political encouragement for cleverer water use might postpone severe water problems. More to the point for the Earth itself, it could allow more lakes and rivers to be left in a comparatively natural state rather than being dammed or encased in concrete.

Fossil fuels and minerals

The damagingly low cost of water is stressed by the International Year of Planet Earth documents. They point out that each year people use over 600 billion tonnes of groundwater alone, leaving aside river and rain water, at a cost of perhaps €300 billion. By contrast, coal and oil production total about 8 billion tonnes – priced at €926 billion.

Oil is the biggest internationally traded commodity, ahead of coffee in second place. In 2006, President George W. Bush claimed that the US is "addicted" to oil, and it is certainly true that all rich societies depend upon it. How severe is their addiction?

In 2005, oil giant BP estimated that the world had about 1190 billion barrels of oil in the ground. That is about 160 billion tonnes. Of that, about 80 million barrels a year are being extracted. While more oil is being found every year, consumption is growing faster, so that the ratio of reserves to production had fallen to just over forty years in 2004.

The picture changes (again using BP's numbers) if you look at coal or gas. There are about 180 trillion cubic metres of **gas** available, equal to 68 years of consumption, although gas use has been growing apace. For **coal**, there are 909 billion tonnes, but we are "only" getting through about 5.5 billion tonnes a year of all types of coal, so we have almost 200 years' worth in the bank.

Move away from fossil fuels into other minerals and the picture is less urgent. Although I have never heard of any nation invading a foreign country to get access to its gravel, the figures show that world use of sand and gravel totals about 18 billion tonnes a year. Much of the **gravel** that is used was made by glaciers scraping Europe, Russian Asia and North America during the last ice age and we will have to wait for the next one for it to be replaced. **Sand**, by contrast, is made every day by the erosion of rock. Because they are the staples of the building trade, their use dwarfs that of all other minerals. The remainder add up to about a billion tonnes a year, of which **iron ore** accounts for about two-thirds.

The difference between using iron ore and using oil is that iron can be recycled, as the big world trade in scrap shows, while a hydrocarbon that is burnt cannot be burnt again. With solid minerals, the real issue is the environmental damage caused by extracting them from the Earth and transporting them around the world – a problem predicted to grow as billions of people in China and India start to demand similar levels of consumption to those seen in the developed world.

Life after oil

No business would take a look in its warehouse and panic because it had only forty years' supply of nails or envelopes left. Why should humans as a species behave differently when it comes to oil?

The main reason is that they are not making oil, coal or gas any more. They are still being produced inside the Earth, but only at a minute rate. Worse, our dependence is so complete that it is hard to imagine an advanced society which does not involve fossil fuels being consumed in huge amounts. But like it or not, alternatives need to be found – and the issue of climate change means we ought to be making the change even before the oil runs out.

At the moment, technology is moving so fast that it seems like hubris for governments to try to tell us where our energy supplies will be coming from several decades from now. In fact, the only idea they seem to have come up with is the Back to the Future notion of building more **nuclear power** stations.

This approach has a number of drawbacks. One is that it produces waste that nobody has really thought how to cope with. Another is that nuclear power stations only make electricity, and most energy is not used in that form. They also take too long to build, the public hate them, and they need vast state subsidies even in such capitalist nations as the US.

But the real reason to oppose them is that they divert skilled scientists and engineers from more interesting opportunities. What would happen if the people who might be used to design, build and run them were told instead to develop machines, vehicles, houses and other devices that absorb no resources and emit no pollution? They might produce bigger, better and cheaper **solar panels**, **wind generators** and the like. But they might also produce surprises we have not thought of yet. Energy can in theory be extracted from any thermodynamic imbalance in nature, even tiny ones. It might be possible for future computers or cellphones to run off tiny air movements or temperature changes. That might not sound like a way to save the world. But it would chip away at energy use and at the same time reduce the sheer weight of power stations, power lines, transformers and the rest that today's society needs to keep things going.

Climate change

Oil, coal and gas may be set to run out in the next century or so, but should we be using them in the first place? People have always known that fossil fuels present problems. Oil causes environmental damage when it leaks from pipelines or ships. Burning it produces a wide range of pollutants. The same goes for coal. In the 1960s, the Central Electricity Generating Board in England decided to avoid complaints about pollution from big coal-burning power stations by building the highest chimneys it could manage to send sulphur dioxide and other combustion products a long way away. The result was an ecological crisis in Scandinavia, with rivers and lakes poisoned by acid rain (see www.acidrain.org). And because most oil reserves are in the Middle East, political problems beyond the scope of this book are bound to follow if whole Western societies depend on it.

In recent years, these long-familiar problems have been overtaken in popular, political and scientific discourse by the issue of **climate change**. The Swedish chemist Svante August Arrhenius (1859–1927) suggested about a hundred years ago that all the coal being burnt might make the Earth warmer by upping the amount of carbon dioxide in the atmosphere. Since then, our knowledge has moved on, but his insight remains valid.

The greenhouse effect

Arrhenius was talking about the **greenhouse effect**. The term is thrown about a lot despite some doubts that the effect is responsible for keeping actual greenhouses warm.

The greenhouse effect works because most of the Sun's energy that arrives at the Earth is in the form of visible and infrared light. That is because the Sun is brightest in that part of the spectrum, as we have seen, which is why it looks yellow. With some exceptions, such as the potentially harmful ultraviolet light that is filtered out by the ozone layer, it passes through the atmosphere and reaches the surface. But the Earth is far cooler than the Sun. So when it emits that energy again, it puts it out at longer wavelengths, deeper into the infrared. (Any physics textbook will tell you more: look under the Stefan-Boltzmann law.) But it happens that molecules of a number of gases naturally present in the atmosphere absorb radiation at these wavelengths, essentially because the chemical bonds between the atoms that make them up resonate at these frequencies. They include water vapour, methane and carbon dioxide. Eventually, the molecules give up the energy they absorb in this way. But when they do so, they emit the radiation that would otherwise have headed into space in any direction at random. So instead of going away, a share of that heat stays behind, and the Earth gets hotter.

We know that the main greenhouse gas in the atmosphere is increasing in abundance. Since 1959, the concentration of **carbon dioxide** in the atmosphere has been measured at Mauna Loa in Hawaii, a remote spot chosen because it is far away from any local pollution (see www.cmdl.noaa.gov). Between 1959 and 2006 the annual average amount of carbon dioxide in the atmosphere rose pretty consistently from about 315 to about 380 parts per million. Parts per million don't sound like much, but carbon dioxide is a powerful greenhouse gas. The greenhouse effect is real physics, so you cannot have this increase without its having an effect.

While carbon dioxide is the main greenhouse gas, the greenhouse effect is self-feeding. Most of the Earth's surface is water. The hotter things get, the more sea water turns to **water vapour**, which is a greenhouse gas itself.

Of course, things are not that simple. The non-believers in global warming point out that more water vapour should mean more clouds, which in turn should reflect sunlight away and reduce temperatures. They add that industrial activity such as coal-burning produces "aerosols" – atmospheric clouds of tiny particles of pollution – that would also reflect energy back into space. Equally, some of the carbon entering the atmosphere may turn into more luxuriant plant life, or more limestone deposits in the oceans.

Maybe the best way to find out whether the Earth is getting warmer is to take its temperature. In the US, the National Climate Data Center (www.ncdc.noaa.gov) has tried to produce average Earth temperatures for the land, the ocean, and the two combined, for the period from 1880 to the present. The figures indicate an average temperature rise of about 1.4°C over that period for the land, and rather less for the oceans. Data gathered through the twentieth century showed that pretty well everywhere on Earth got warmer during those hundred years.

Of course, the Earth is part of a more complex system. It has been suggested that the warming that is being observed is to do with the Sun putting out more energy, or with the Earth's orbit allowing it to receive more. However, the majority of scientists now agree that artificial "anthropogenic" emissions of greenhouse gases are the primary cause of global warming.

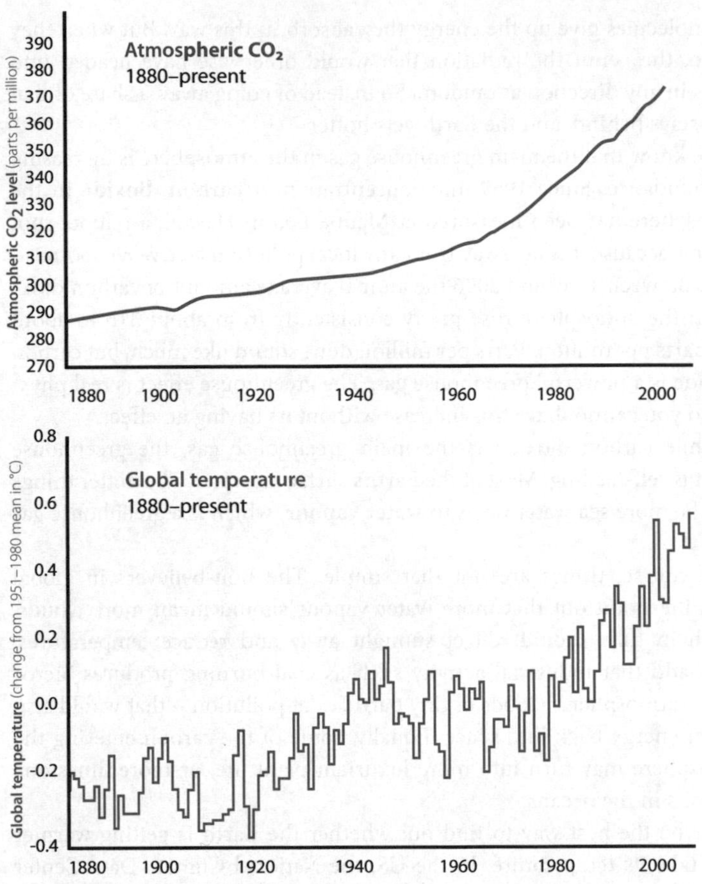

The results of global warming

Global warming is about climate and weather. Both of these are complex systems that are best discussed statistically, not in terms of specific events. You cannot point out of the window at a hot day and claim that it is due to global warming. Equally, even in a warming world there will still be unexpectedly cold winters – and summers. This applies especially to extreme weather. A single severe storm cannot prove global warming is happening. However, when New Orleans was devastated by **Hurricane Katrina** in 2005, many experts agreed that this was a rare example of a storm that would probably not have happened, at least on the scale it did, before the era of global warming.

Thinking about what a warmer world would look like has become a major intellectual industry. Most of the prognostications turn out to be unhappy ones.

As we saw in the previous chapter, there would likely be significant **sea level rise** as major ice masses melted in Greenland and the Antarctic, as well as more regional change such as altered flows in the world's major rivers caused by melting glaciers. Another unavoidable big effect would be increased **erosion** and **storm damage** on ocean shorelines, including ports and cities as well as beaches. Less certain, but looking very possible, would be a large increase in **violent storms**. There have always been storms, and some enemies of the idea of global warming claim there are no more now than there ever were. However, research published in 2005 suggested that although there were about as many hurricanes as before, the number of them that were very severe was rising. The scientist in charge, Peter Webster, said that rising sea temperatures might well be responsible. If there are to be more violent storms, the southern US and Caribbean will be most affected. But they would push further north in the US and also become more common in other storm-prone countries such as Indonesia and Australia.

The Earth's zones of vegetation cannot just move further away from the Equator as warming occurs: **deserts** would spread and **farming** would be disrupted. And because a warmer world is also wetter, many parts of the

New Orleans: water finally gets the better of engineering in 2005's Hurricane Katrina disaster

Climate change deniers

What should we do with our growing awareness that we are altering the Earth's climate? Some think that innovation, especially in energy technology, will sort it all out, given some political will and an awareness that there is profit in it. This is more or less this author's line, although I also think that time is short and we ought to get going with more enthusiasm than we are showing right now. Another is to assume that we are helpless in the face of catastrophe brought on by the malign influence of international capital. This is a counsel of despair and is also untrue. But the most puzzling response to climate change is to deny its existence.

The terms of the debate have shifted radically in recent years. The minority of scientists who do not accept the general framework of global warming has shrunk. The firms that once resisted the idea in case it began eating into their profits have mostly realized that a change to new energy sources means new business for them. After all, it seems that we have used just about half of the Earth's readily accessible oil, so oil companies need to think long-term about diversification in any case. And at the same time, the effects of climate change, especially the general acceptance that it had a hand in the 2005 flooding of New Orleans, have got harder to ignore, and so have the costs. This has meant that a series of climate change denial organizations such as the Global Climate Coalition have lost credibility as well as funding from oil companies and others. Politicians may not do much about climate change but the era is probably over when they could claim it was a myth.

However, climate change denial lives on in a few places. Some scientists and enthusiasts like to claim that every unknown in the equations proves something is horribly wrong with our ideas, or insist that every new discovery shows that what we knew before was wrong. These folk are prone to one massive form of inconsistency, though. They often seize with glee on reports that climate change is a good thing. Although such change would be very damaging to the developing world, for example, there are occasional reports that it might have positive effects, such as making crops grow faster. Often the very people who say climate change is not occurring are the ones hyping up reports of this kind claiming that it is a good thing.

world would have **cloudier summers**. So global warming might wreck the skiing, but not compensate by providing several months a year of beach parties. In addition, a 2006 study for the UK government showed that human **diseases** now present in Africa, such as malaria, and animal ones such as Bluetongue, a devastating cattle disease, might spread north into Europe, with direct effects on human health and well-being.

The real problem with global warming is that it is not happening in a wild world whose natural systems will sort things out. When Europe warmed up in the Middle Ages, the effects were limited because most of the affected area was wild. When the climate changes now, things are very different. The obvious case is sea level rise. Sea levels around the world rose by 10–25cm during the twentieth century. When the sea rises, the coast does not just move back. Instead, it arrives at houses, ports and concrete sea defences. The narrow strip of beach, mangrove or some other wild land is the first to be destroyed.

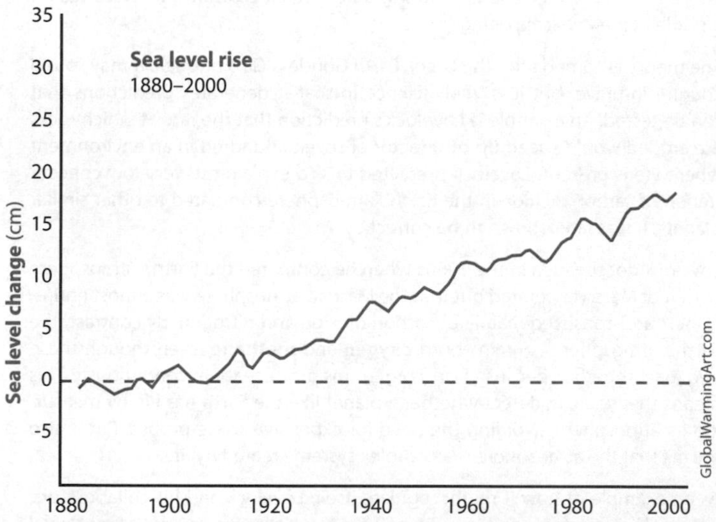

As we have seen, the root cause of global warming is the fact that carbon built up inside the Earth over millions of years is being released in just a few decades. It is not surprising that this has environmental effects. But it also means that doing something about it is tricky. A raft of steps have been proposed to cut fossil-fuel use, or to reabsorb the carbon they release. The measures suggested range from expanding nuclear power to planting more trees. But it is not likely that carbon dioxide concentrations will start to come down before the middle of the century. That means that temperatures will probably go on rising at least until then.

The immense effects of global warming described above would far exceed all the previous environmental damage done by humans. They are also genuinely global, while most environmental problems are local or national in scale. So let's think of global warming as a huge issue in its own right – but also as a cue to reorganize the whole way we deal with the world around us.

Gaia

Since the 1970s a group of scientists centred on **James Lovelock**, a British chemist specializing in changes in the atmosphere, have pushed the **Gaia theory**, whose message is that the Earth, far from being a jumble of interacting components, is a self-regulating system that might be regarded as a single organism.

Of course, Gaia would be completely unlike all the other living creatures we know. They reproduce and live among other similar creatures, whereas Gaia is a single planetary-scale being.

The theory is named after the Greek Earth Goddess Gaia or Gaea. It may sound counter-intuitive, but it is "real" science in that it generates predictions that can be tested. An example is Lovelock's prediction that the rate at which rocks are eroded would speed up by a factor of several hundred in an environment where life is present. Lovelock predicted this to explain the very low concentration of carbon dioxide in the Earth's atmosphere compared to other similar planets. It was later shown to be correct.

Lovelock got the idea in the 1960s when he compared the Earth's atmosphere to that of Mars. He pointed out that the Martian atmosphere was almost entirely inert and consisted mainly of carbon dioxide and nitrogen. By contrast, the Earth's atmosphere contains both **oxygen** and **methane**, even though these two react together, because living organisms produce them continuously. This means that you can detect whether a planet like the Earth has life by measuring its atmosphere, avoiding the need for expensive space probes. But it also means that the atmosphere is a complex system created by life.

As an example of how it might regulate itself, Lovelock and his collaborators, notably US biologist Lynn Margulis, point to **temperature**, one of the most vital aspects of Earth as a home for life. One of the main influences on the Earth's temperature is how cloudy it is. Clouds are white and if you have more of them, the Earth will reflect more solar energy back into space, upping its albedo (see p.218). Lovelock pointed out that a chemical emitted by algae called dimethyl sulphide encourages clouds to form, by creating aerosols of small particles on which water vapour can condense. The hotter things get, the more the algae thrive and the more dimethyl sulphide is created. This means more clouds, and eventually a cooler Earth. But if the Earth cools, there are fewer algae, the clouds dwindle and the temperature goes back up. So the whole system is self-adjusting to keep the temperature level.

You could argue that there is no intention in any of this. It just happens. But then, primitive mammals never intended to evolve into you and me. Over time, Lovelock argues, it is in the interest of life on Earth to develop systems that keep things comparatively stable. Even ice ages have a role in Gaia, and stop and start when the Earth gets too hot or cold.

One theory is that this planetary control system has developed over time to keep things stable in response to a gradual increase in the Sun's energy output over billions of years.

But it is not Lovelock's message that we can do anything we like with the Earth and wait for Gaia to put it right again. The planetary management system he describes works to cope with change on a geological timescale. It risks being overwhelmed if we push vast amounts of greenhouse gases into the atmosphere, or remove a large part of the tropical jungles, much as it takes a long time to recover from a major volcanic outbreak or the sort of meteorite strike credited with the mass extinction that killed off the dinosaurs.

Some think that Gaia will cope with this crisis by getting rid of us, or drastically reducing our numbers. This is the fate with which Lovelock threatens us in his latest book, *The Revenge Of Gaia*. But even Lovelock accepts that there may still be time to avoid this crisis by limiting population growth, overuse of the land and the oceans, and our consumption of fossil fuels.

Gaia theory sees the whole planet as a single organism

Extinction

"Extinction is for ever" used to be a green movement slogan. It is no longer completely true. Thanks to the wonders of genetic engineering, it is becoming possible to recreate lost creatures provided we have their DNA. The quagga, a species of zebra whose last member died in Amsterdam Zoo in 1883, is one animal that might get the recreation treatment soon (see www.scienceinafrica.co.za/2003/july/quagga.htm).

In case you are wondering, it is far trickier to do this with long-extinct animals with no living relatives, for example dinosaurs. One reason is that the plants they used to eat have also gone extinct and those that exist today might be toxic to them. More importantly, you need a near relative to give bith to the revived species.

Is extinction for ever? The last quagga died in Amsterdam zoo in 1883 but its preserved DNA may allow it to be cloned via other types of zebra.

Although some creatures can be driven to extinction by humans killing them off, far more are put in danger by having their habitat removed around them. So like climate change, extinction is important in its own right but also for what it tells us about how we are changing the Earth.

Just how many species are in danger of being wiped out isn't an easy question to answer. Our knowledge of the world around us is not perfect. The excellent Recently Extinct Animals website (www.petermaas.nl/extinct) run by Peter Maas in the Netherlands lists dozens of species, such as the Gomeran lizard (in the Canary Islands) and the Bavarian pine vole, that have been spotted alive and well some time after reports had been issued of their dying out.

The International Union for Conservation of Nature and Natural Resources runs the Red List of endangered species (www.redlist.org). It turns out that the bigger and more noticeable a species is, the likelier biologists are to know about it in detail. Thus there are 5416 known species of **mammal** of which we have detailed knowledge of 4853. Of these, according to IUCN, 1101 were threatened with extinction in 2004. That is a startling number, almost exactly a fifth of the total. For **birds** the figure is 12 percent, at 1213 out of 9917 species to have been described; for **amphibians** it is an even more alarming 31 percent, 1770 out of 5743 described species.

However, things get a lot less clear when less adorable species are being considered. For example, there are about 950,000 described species of **insect**. Of these, IUCN was only able to evaluate the status of 771 properly for its 2004 return. It turned out that 569 of these were endangered. That is 73 percent of those evaluated, but less than 0.1 percent of the total number we know of. Although the species chosen for analysis were probably the most imperilled, it is likely that the real number of threatened species is far higher than the headline figure. The same goes for the **lichens**, creatures made up by the co-operative union of fungi and algae. There are said to be 10,000 species of lichens of which two are in danger, but in fact these two are the only ones to have been properly investigated.

The IUCN's tables show that the numbers of species apparently at risk have been increasing consistently since the mid-1990s. Perhaps most alarming is the fact that the list contains over 8000 species of **plants**. While people might hunt bears or wolves to death, the extinction of a plant usually stems from gross alteration of the environment, and in turn threatens the animals that eat it. The number of endangered plant species counted by IUCN has risen by about half since 1996. The figure is slight compared to the number of plants known – 287,655 species, to be exact – but also adds up to 70 percent of the species studied in detail.

The IUCN's headline figures suggest that only about 1 percent of species are threatened with extinction. However, a far higher percentage of the species that have been studied in detail are in trouble. The figures support an estimate that extinction is now running at anything up to 100 times its normal rate.

What are we doing to cause all this extinction? Some species, particularly large land mammals, have been deliberately hunted to the point of extinction. Now, tigers and elephants are seen as having value in their own right, if only as something for lucrative tourists to come and look at, which may help them survive. Others, such as whales, have been hunted for food, although here, again, there has been an improvement in recent years despite some nations' continued insistence on hunting whales. The southern right whale population plummeted to about 300 in 1935, but has now grown back to about 7000 members with the cessation of most hunting. A similar story can be told about a number of bird species.

Most of the species that are in danger of being wiped out directly by human activity are **fish**, because demand for fish, and fishing technology, have both developed apace in recent decades. Climate change has been blamed for severe declines in some other species such as frogs and toads in Costa Rica, reduced radically in numbers by unusually long periods of drought.

But as we saw above, the greatest cause of extinction is damage to a species' native **habitat**. The IUCN points out that up to half the world's freshwater turtles and a third of its amphibians are threatened with extinction, a fact which reflects the huge stress that pollution, agriculture and development are placing on the lakes, rivers and wetlands that they need.

Quite simply, it is the sheer percentage of the Earth's biological activity that we are making use of that is sending species extinct, as we concrete the river banks and plough up the jungles. Although it is unlikely that we'll be responsible for a mass extinction on the scale of the one which saw off the dinosaurs (see box), the prospect is a gloomy one. Some espe-

Are we causing a mass extinction?

Are we really living through a mass extinction? If so, we ought to take care. The lesson of the past is that the biggest and most dominant creatures are most likely to vanish when such a mass extinction gets going, while the more modest ones keep plugging on. If humans are using up a large percentage of the Earth's resources, they are more vulnerable, not less, to big changes in the Earth system.

There have been estimates that the rate of extinction is now as much as 100–1000 times as high as it would "normally" be. But is that the same as a mass extinction? There is room for doubt about the figures. We know a lot about past extinction from the fossil record but our knowledge of today's living species is in some ways less complete.

The most famous mass extinction took place at the end of the **Cretaceous**. Although it is mainly remembered for the demise of the **dinosaurs**, the destruction extended to fish and many other sorts of plant and animal. Before the **asteroid impact** theory of their extinction became popular, there was a whole industry devoted to thinking about what killed the dinosaurs, with some spectacular ideas such as mass constipation. But none of these theories ever explained how the same cause killed off all the other species as well. Even the ammonites, a hardy family of swimming marine animals that had survived previous mass extinctions, died out.

The extinction at the end of the Cretaceous involved the destruction of about 85 percent of species in the fossil record. On the basis of this event, we are certainly

cially shroud-waving experts claim that up to a fifth of the Earth's species could vanish within thirty years.

Seeing the future

Some thinkers have already decided that we have made such a mess of the Earth that our future lies elsewhere. They are wrong for many reasons, not least because the solution they propose is far too complex. We do not know of any other place in the universe remotely as nice for humans as this one. There is everything to be gained by curing the problems we have created, not running away from them. In a million years, it may well be that most people will not be living on the Earth, but for decades to come, they certainly will be.

As we have seen, the global population looks likely to rise by at least another third. But the Earth can support a population of this size with rational technology used in a sustainable way. The question is not how

---

not living through a mass extinction today. There is little chance that human activity as we see it now will wipe out 85 percent of species. If it did, *Homo sapiens* would probably be among them, which would solve the problem.

But one thing we know about these extinctions is that they are not simple events in which species curl up and die en masse. As geology has got cleverer, we have been able to see that some species die out many thousands of years after others.

This seems to make sense because of everything we know about ecosystems. We are now aware that plants, animals and other types of life form subtle networks in which members depend upon each other.

Although a meteorite strike that chills the Earth overnight and kills most of the life it contains is one image of a mass extinction, there are others. At the end of the **Permian**, 95 percent of the shallow marine species went extinct, but there is no sign of a meteorite impact to explain what happened. It may have had more to do with plate tectonics and sudden changes in the size and shape of the oceans.

So although green propagandists are wrong to say that we are living through a mass extinction as the movies portray them, maybe we are at the start of something like a mass extinction as they more usually happen. Here, props are slowly knocked away which will eventually lead to the structure of life on Earth falling in.

numerous we might be in the future but what we will be doing. If we are clever about how we use the Earth and its resources there is no reason why there can't be a future for us here.

The good news is that we are in a position to solve the problems that lie ahead of us. As we have seen in these pages, we now understand the Earth in a very complete and satisfying way. Just think: today we know that the Atlantic is widening at a few centimetres a year and can use laser beams bounced off satellites to measure it as it does so. In the Earth's scheme of things it is only a tick of the clock since Columbus set out across that same ocean under a complete misunderstanding of how large the Earth was – hence his belief that he had got to the Indies when in fact he had reached the Caribbean. We can now model with exquisite accuracy parts of the Earth's interior that we have no way of reaching, and can predict the climate, this time with less accuracy but with a good idea of where the problems in our forecasts lie. Equally, our ability to explore other planets and to see our own with godlike gaze from space has transformed the way we think about the Earth.

At the same time as we are gaining all this knowledge, people all over the world are getting better educated and more able to understand the choices ahead of us. And the spread of the web – which is only in its infancy as a source of information – means that more people can find out more up-to-date facts than ever on which to base their opinions and reactions. Our ambition as a species may have brought about many of the problems our planet now faces, but it will also help us to solve them.

9

The future Earth

The future Earth

Scottish geologist James Hutton, one of the founders of modern Earth science, said in the 1790s that geological investigation showed "no vestige of a beginning, no sign of an end". But knowledge has moved on since then. The beginning is now clear enough. We know how the Earth formed and when, about 4.54 billion years ago. And we even have a fair idea of its life expectancy.

Although there are many possible disasters that could affect conditions on the Earth (see boxes on pp.266–67 and p.268), there are few we know of that could affect its very existence. It is likely that the end of the world as we know it will be brought about by the future activity of the **Sun**. As we have seen, the Sun is already hotter than it was a few hundred million years ago. In perhaps 5 billion years, it will be so much hotter that it will have boiled away the oceans, making the Earth essentially as comfortable for life as Mars is today, albeit a lot warmer. As we have seen, the Gaia theory holds that the Earth can cope with major disruptions provided they are gradual enough. But it may be that this increase in the Earth's temperature is beyond even Gaia's powers of adaptation. Even if it can cope, there is worse to come.

In about 6.5 billion years, the Sun will have finished the **hydrogen burning** that provides most of its luminosity today. Instead, it will stop being a "main sequence" star and will run through a scrics of comparatively rapid changes over a few hundred million years.

In the first of these, the Sun will swell to a size which will engulf Mercury, and raise the Earth's surface temperature to perhaps 750°C. Later, it will swell to about the size of the Earth's orbit today. This sounds terminal, but in fact it will not be. Because the Sun will have blown off large amounts of its mass into space, the Earth's orbit will have expanded as the Sun got lighter. Finally, the Sun will erupt to produce a gas cloud called a **planetary nebula** (because it can look like a planet when seen in a telescope) and cease almost all heat production. At this stage the ashy

Asteroid hazard

Ceres, the biggest asteroid, was discovered on the first day of the nineteenth century. It is about 1000km in diameter and since 2006 has been officially promoted to the status of Dwarf Planet. Now we know over 15,000 asteroids, and in 2006, 831 of them had been classed as potential hazards. This essentially means that they have orbits that bring them critically close to the Earth, and are more than 1km across. At the time of writing in 2006, the most significant asteroid hazard known was the possibility of the 320m asteroid Apophis striking the Earth on 13 April 2036. NASA and others are already working on countermeasures. This impact would cause severe damage but would not threaten mass extinction.

By contrast to the millions who have been killed by flood, earthquake and volcano, there seem to be few if any cases on record of a meteorite or asteroid killing anyone, although people have been struck by meteorites. But although asteroid impacts are rare, they are inevitably very damaging. Even a small one could cause as much destruction as a nuclear weapon, while a large one could mean **mass extinction**.

Perhaps the most remarkable fact about an asteroid impact is that, with a little new technology, it is the most preventable of natural disasters. It is tricky even in principle to do much about a volcano. It may become possible to predict earthquakes more reliably than we do today, and to design buildings that withstand them better, but stopping one would require godlike abilities. Likewise, complete control of the climate remains in the realm of science fiction.

By contrast, our awareness of asteroid hazards has been piqued by the fact that we can contemplate doing something about them. First, we have enough telescopes and cameras around the Earth to allow "Near Earth Objects" to be spotted more reliably than ever. One could still arrive from a clear sky, but systems now in place are diminishing the chance of such a surprise. In addition,

leftover of the Earth will orbit the remains of the Sun for ever, lit only by distant starlight and with its temperature far below freezing.

Even further in the future lie such possible fates as the **heat death** of the universe, in which it has "run down" to a state where there is no free energy to sustain motion or life, or the **Big Crunch**, when galaxies merge back together as the expansion of the universe goes into reverse.

In the meantime...

Our knowledge of stellar evolution, the development of stars over time, is solid enough to make the events described above pretty definite. But the any impact that was predicted would probably turn out to be decades in the future. So humanity would have plenty of notice.

What to do during the warning period would be a trickier choice. One idea would be to blow the asteroid to bits with nuclear weapons. However, this means people having atomic bombs, a far worse hazard to the species than any asteroid. Also, a substantial percentage of the bits of the asteroid would probably continue in much the same orbit as before and hit the Earth, with lesser but still damaging effect.

But there is no dearth of more graceful ideas. One would be to place a tiny rocket motor on the asteroid and gently nudge it into an orbit that misses the Earth. Another would be to paint one side white so that the different pressure of sunlight on its faces would shift it in its orbit, although as asteroids rotate, this might not be feasible.

While asteroid hazards to the people of Earth have not been increasing, the attention paid to them has grown rapidly. Doubtless some of the running has been made by an aerospace industry that smells profit. But if they can make money preventing our demise, why not?

Perhaps we need more creative thinking about what to do with a possible impacting asteroid. An obvious idea would be to divert it into orbit round the Earth. Even a tiny one would contain raw materials for any number of space stations or planetary probes. But perhaps it would be more fun to send it full-tilt into the near side of the Moon so we could see a new crater form before our eyes, and marvel at our cleverness in being at a safe distance as it happened.

For more information see:

Asteroid and Comet Impact Hazards impact.arc.nasa.gov/intro.cfm

many billions of years before any of this happens contain fewer certainties. What we can say for sure, however, is that nobody will be killed or even inconvenienced when the Sun boils the oceans and fries the Earth. By then, humans will either have gone extinct or spread across the galaxy, depending on your level of optimism.

Working from the inside out, the innermost layers of the Earth are perhaps the most likely to behave as we expect for many billions of years to come. Here, the process of differentiation which we met at the end of Chapter 4 will continue to operate. Over the last billion or so years, the solid **inner core** of the Earth has grown from nothing to some 2444km in diameter. It will carry on growing at about 0.3mm a year as the core loses

Supernova alert

A **supernova** is a star which has erupted on a terrific scale. When we see one in a distant galaxy, it can outshine all the other millions of stars that the galaxy contains. Our own Sun is not massive enough to go supernova, but that does not mean we are safe from them. One could still go off next door, on a cosmic scale of distance.

While experts do not agree exactly, it seems that a supernova within perhaps 100–200 light years of the Earth would have severe effects. The blast of x-rays, gamma rays and other **cosmic radiation** that it would produce would be most damaging. It could strip the ozone layer (see p.132) that protects us from the Sun's ultraviolet light. Then solar radiation would kill off some species, while depriving the rest of food and speeding up their rate of genetic mutation.

There are so few stars within 200 light years of the Earth that this is not a short-term risk we need fuss about. There is one possible future supernova, **HR8210**, only about 150 light years off. But even its possible explosion is hundreds of millions of years in the future.

However, the age of the Earth makes it pretty certain that there have been nearby supernovae within its lifetime. Heavy metal traces in rocks associated with the mass extinction at the end of the **Ordovician**, 440 million years ago, have been linked to a possible supernova. This extinction killed most extant species and was the most severe we know of. And the amount of time available means that there may well be another nearby supernova to attack life on Earth before the swelling Sun boils the oceans.

heat to the crust and eventually into space. At some point, it could fill up the whole of the volume of today's core.

Likewise, the **geodynamo** is certain to go on doing what it does for many millions of years. As we have seen, one of its most lasting characteristics is its tendency to flip by 180° from time to time, perhaps under the influence of small perturbations caused by the growth of the inner core, irregularities in the base of the mantle, or even asteroid impact. Whatever the root cause, the switch seems to happen when the Earth's magnetic field falls to about 10 percent of its peak value. As we have seen, some scientists suspect that this could be soon, partly because the magnetic field is weakening measurably and partly because it has been a long time since the last flip. However, even if the next flip is a long time in the future by our standards, many thousands more reversals lie in the Earth's future.

The fact that the inside of the Earth is hot means that like the core, the Earth's **mantle** will carry on convecting indefinitely. This means that there will be **drifting continents**, and the volcanoes and earthquakes that go with them. But where?

One person who has attempted to answer this question is US geologist Chris Scotese who, as well as producing maps of past continental drift, has run the clock forward by millions of years. Even after 50 million years, things look noticeably different, with the Mediterranean closing up. By 250 million years in the future, the ancient continent of Pangea has more or less reassembled itself, with all the land in one large mass. (See maps overleaf.)

This continent will be quite unlike anything we know. Most places will be a long way from the sea. Big rivers will cross the land mass, but substantial estuaries and deltas will occur only around the edge. There will be seafloor spreading in the continuous deep ocean surrounding the land, and subduction around the edge of the continent, within which there will be a big trapped sea. All that subduction will mean lively volcanoes and earthquakes. (Scotese is careful to say that a forecast this far ahead cannot be completely dependable.)

The future world that Scotese foresees would differ from the one we know in a number of ways. One is that it has no large land mass at either pole. This means that there would be no Antarctic ice cap because there would be no Antarctic continent. But there would also be no frozen ocean like the one we now find in the Arctic. This icy sea exists as it does because it is more or less cut off from the rest of the world ocean by the American and Eurasian land masses. If there is one large land mass, ocean currents, ice distribution and even atmospheric circulation will all be different too.

In the shorter term, one of the greatest uncertainties about the Earth's future is its **climate**. We already know that we are (as former British Prime Minister Margaret Thatcher put it) carrying out an experiment with the Earth, by releasing carbon that was stored in the crust over millions of years during a conflagration lasting just centuries.

As we saw in Chapter 8, global warming is likely to make life on Earth less comfortable for humans over the coming centuries. But what about further down the line? Although the precise details are unclear, Tim Lenton of the University of East Anglia in the UK has looked at how stable the Earth is for the life it bears, with an emphasis on water, air and land rather than the depths of the solid Earth. He points out that something not too different from today's unplanned climate experiment happened about 55 million years ago. Then, large amounts of carbon were released into the atmosphere in the form of methane. The release seems to have happened because compounds called clathrates, in which methane is trapped in ice, were melted by a comparatively small temperature rise.

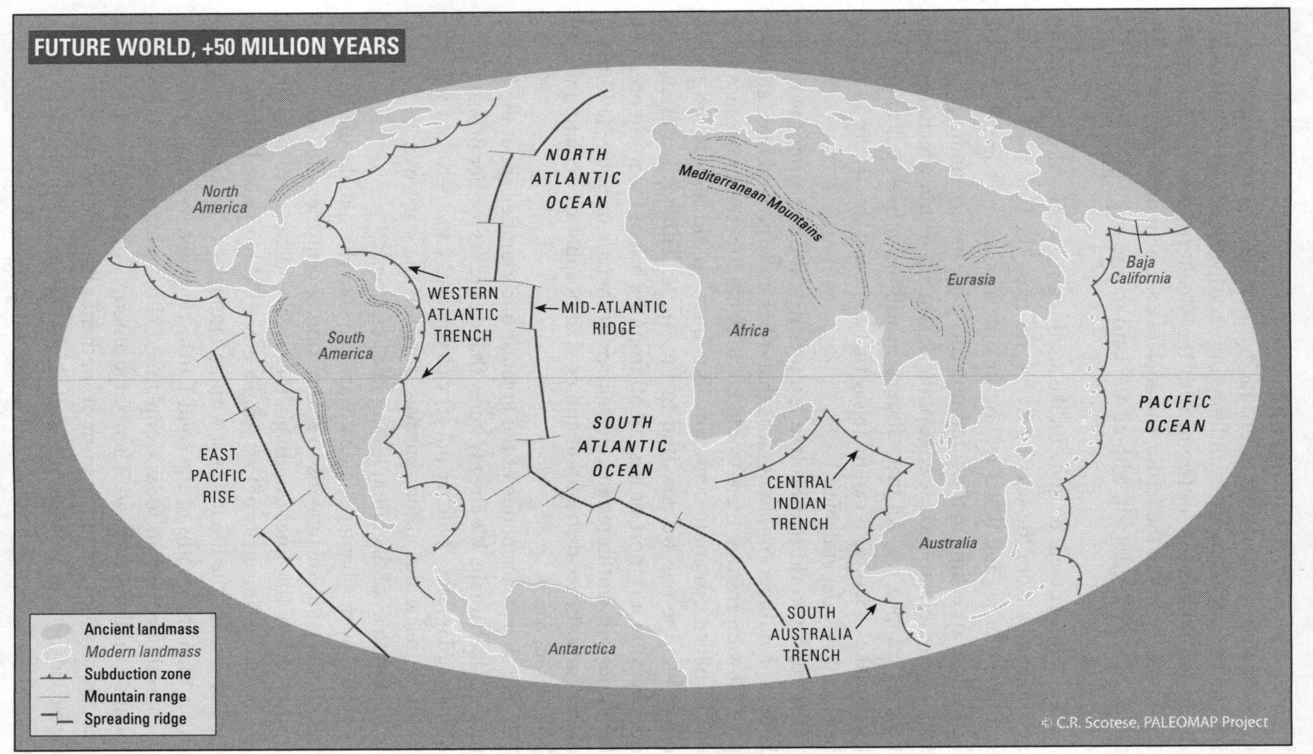
FUTURE WORLD, +50 MILLION YEARS
North America
NORTH ATLANTIC OCEAN
Mediterranean Mountains
Eurasia
Baja California
WESTERN ATLANTIC TRENCH
MID-ATLANTIC RIDGE
Africa
South America
PACIFIC OCEAN
SOUTH ATLANTIC OCEAN
EAST PACIFIC RISE
CENTRAL INDIAN TRENCH
Australia
SOUTH AUSTRALIA TRENCH
Antarctica
Ancient landmass
Modern landmass
Subduction zone
Mountain range
Spreading ridge
© C.R. Scotese, PALEOMAP Project

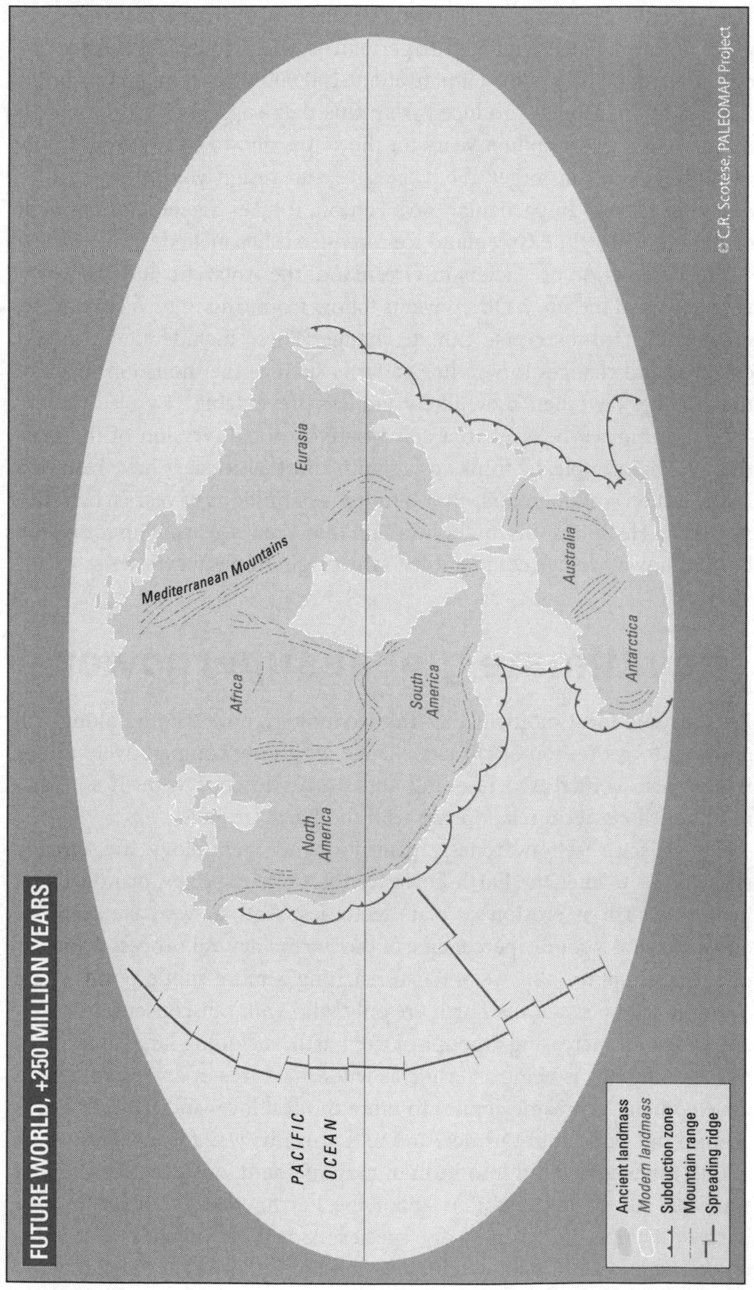
FUTURE WORLD, +250 MILLION YEARS
PACIFIC
OCEAN
Mediterranean Mountains
Eurasia
Africa
South America
North America
Australia
Antarctica
Ancient landmass
Modern landmass
Subduction zone
Mountain range
Spreading ridge
© C.R. Scotese, PALEOMAP Project

The effect was probably similar to the release of methane that may take place in future in the Arctic if the permafrost melts (see p.217). After this episode, it took the Earth's climate about 100,000 years to regain its original condition. Models produced from this data suggest that this time, it might take about a million years for the status quo to be restored. In the meantime, you can forget about ice ages and resign yourself instead to higher sea levels. In particular, says Lenton, it takes a geological amount of time to regrow the Greenland ice cap once it has melted.

The destruction of glaciers in Greenland, the Antarctic and elsewhere is only part of the story. Other events follow from, and in turn encourage, major and hard-to-reverse climate change. These include altered ocean currents and changes to weather patterns such as the monsoon. Each of these makes the others more likely and less preventable.

Lenton himself believes in a comparatively tough version of the Gaia theory. For example, he thinks it is significant that ice ages have restarted in the last few million years, after hundreds of millions of years when they were rare. He thinks this might be a sign that Gaia is gearing up for major, but unknown, change comparable to the origin of photosynthesis.

The ultimate global superpower

Let's assume for a moment that the technology now coming along will allow us to get on top of climate change with only comparatively minor issues such as increased flooding and storms to cope with. If so, what might our long-term relationship with the Earth be like?

Our species' self-awareness, planning and technology are already allowing us to alter the Earth far faster than other species, or non-living processes such as erosion or plate tectonics. And, as we have seen, we already corral a goodly percentage of the Earth's natural processes for our own consumption. But we are also reaching a more subtle point where even the wild parts of the Earth are only wild with our consent. It is welcome that an increasing amount of the Earth, including large areas such as the Antarctic, is being set aside as wilderness free of extensive human development. The same applies to more modest local and national parks around the world. But our decision to act in this way means that we are taking the whole world into **human management**, ensuring for the first time that all the big decisions about the Earth, even decisions to leave parts of it alone, are taken by one species. As Bill McKibben puts it in his book *The End Of Nature*, we could let human activity alter the climate so that a hotter-than-before day is essentially a human artefact. But if we take steps to stop altering the climate, a normal cool day is also an artificial event which we have decided to create.

Some Gaia theory enthusiasts think that we are now damaging the Earth so much that it might decide to do without us, and drive us to extinction. But James Lovelock, the originator of the idea, disagrees. His view is that our self-awareness and the way we view the Earth as a whole adds to Gaia's abilities and makes us vital to its future.

If our role is becoming that of planetary "overseer", managing the Earth as a whole for our own ends and at the same time preserving it for other species, how might this role develop further in the future? One intriguing series of possibilities is raised by the Russian space scientist **Nikolai Kardashev**, who has suggested a three-fold typology of civilizations known as the Kardashev scale. We have not yet got to **Type I**, in which all the energy arriving at a planet is under control, so that (say) the weather can be managed accurately. But tens of thousands of years beyond that comes a **Type II** civilization, whose members have control of the whole energy output of a star. **Type III** (the last one he suggests) would control the energy output of an entire galaxy of billions of stars, a form of civilization so mind-numbing that even science fiction has rarely grappled with it. In simple numbers, a Type I civilization would use about 10^{16} watts of power, in other words 1 followed by 16 zeros. At the moment, the energy we use is equivalent to about an hour a year of the solar power we receive. Satisfyingly, Types II and III would get through 10^{26} watts and 10^{36} watts respectively.

The trajectory of human development so far, especially in the last few centuries, may suggest that we are indeed on track for at least Type I status, and Kardashev has speculated that we could reach that stage within 100 years. Observers of contemporary globalization sometimes claim that the emergence of global trade and communications, the arrival of multinational institutions such as the UN, or even the establishment of the Euro, are signs that a coherent Type I civilization may be on its way. However, anyone who sees the problems the UN faces, or observes the stubborn way in which most political power stays in national hands, might not agree. Kardashev's own view is that getting to Type I is the toughest transition. A glance at the TV news suggests he is right, but we have no information on how tricky the next two transitions would be.

If we did attain Type I status, it would be a benign result for the Earth because it would turn humanity into a solar-powered society with few long-term demands on the Earth's own resources. This might seem a far-fetched idea. But one in which history went into reverse and human expectations shrank would be even more surprising.

Whatever change the future holds, one thing we know for sure is that our future is intimately connected with that of our planet. It is no surprise that we find it uniquely beautiful and fascinating. The human race is exquisitely adapted to life on Earth, as a result of more than 3 billion years of evolution here by us and our ancestors, and there is no place like home.

Resources

Resources

There are numerous books, magazines, websites and museums out there which focus on planet Earth. Here are some of the best.

Books

General

The Oxford Dictionary Of Earth Sciences Alisa Allaby and Michael Allaby (Oxford University Press, 2003)
Exhaustive and well written, with many thousands of brief but accurate entries. But be warned: an afternoon can easily be lost browsing its pages.

The Oxford Companion To The Earth Paul Hancock and Brian Skinner (Oxford University Press, 2000)
Eccentrically organized (it has a diagram of a nuclear reactor, but no article on the tides) but accurate and comprehensive once you work out how to use it. Covers the whole Earth, with an emphasis on the solid parts. Worth considering if you want to follow this book up with something more rigorous and scholarly.

The Energy Of Nature E.C. Pielou (Chicago, 2001)
An original look at the Earth from the perspective of the energy that drives it. Pielou takes a tour of energy on Earth, from its arrival from the Sun, to how it causes winds and tides, is captured by living things, manifests itself in magnetism, volcanism and earthquakes, and is harnessed to fuel human society.

Planet Earth

The Compact NASA Atlas Of The Solar System Ronald Greeley and Raymond Batson (Cambridge University Press, 2001)
Drawing on data from numerous NASA missions, this reveals beautifully what we know about the planets and satellites of the solar system. A cross between an atlas and an encyclopedia, it includes detailed information on the geological make-up of each space object.

The Planets David McNab and James Younger (BBC, 1999)
The companion volume to a terrific TV series, this is full of amazing photography and other visuals. It presents with equal excitement the wonders of the solar system and the space missions which have uncovered them.

The solid Earth

The Earth: An Intimate History Richard Fortey (Harper Collins, 2004)
In a class of its own as a literary account of what we know about the solid Earth, this is a beautifully written tale that starts slow and eventually covers just about everything. Also notable is Fortey's *Trilobite!* (Harper Collins, 2001).

The Crucible Of Creation: The Burgess Shale And The Rise Of Animals Simon Conway Morris (Oxford University Press, 1998)
The leading scientist behind our knowledge of the Cambrian explosion of life explores the evolutionary history revealed in "the most wonderful fossil deposit in the world". In doing so, he takes issue with Stephen Jay Gould (see below).

The Evolving Coast Richard A. Davis Jr (Scientific American Library, 1997)
Covers issues such as erosion, deposition, coastal landscape and environmental change as clearly as you'd expect from a product of the *Scientific American* stable.

The Practical Geologist Dougal Dixon and Raymond L. Bernor (Fireside, 1992)
A handy and well-illustrated look at rocks and minerals; also describes the Earth's formation and development.

Wonderful Life Stephen Jay Gould (Hutchinson Radius, 1990)
The Cambrian explosion story from one of the great science popularizers. Compare Conway Morris, above. In addition, almost anything Gould wrote is worth reading.

History of Earth science

The Seashell On The Mountaintop Simon Cutler (Arrow, 2003)
Biography of Steno, author of the law of superposition (see p.93) and arguably the founder of scientific geology.

The Map That Changed The World Simon Winchester (Viking, 2001)
The story of William Smith and the first geological map, a huge, beautiful chart of the strata of England and Wales. Winchester recounts working-class Smith's struggle to be admitted to the scientific establishment.

Compass: A Story Of Exploration And Innovation Ian Gurney (Norton, 2004)
A fascinating history of the magnetic compass, from its first appearance in twelfth-century China to 700 years of attempts to make it easy and reliable to use. Gurney peppers his account with tales of maritime disasters caused by its unreliability.

Latitude And The Magnetic Earth Stephen Pumfrey (Icon Books, 2002)
An account of Elizabethan scientist William Gilberd's discovery of the Earth's magnetic nature and his technique for using the Earth's magnetism to determine the latitude of ships at sea.

The airy Earth

How Weather Works: Understanding The Elements René Chaboud (New Horizons/Thames and Hudson, 1996)
A well-illustrated guide to the workings of the weather, how we find out about it and how our understanding of the weather has changed over time.

Air Apparent: How Meteorologists Learned How To Map, Predict And Dramatise Weather Mark Monmonier (Chicago, 1999)
World expert on how the world has been represented in maps, Monmonier here turns his attention to the weather and how it can be shown on screen and paper.

Atmospheric Change: An Earth Systems Perspective T.E. Graedel and Paul J. Crutzen (W.H. Freeman, 1993)
Surprisingly undated account of the climate system, written by the world's foremost climate experts, Nobel Laureate Paul Crutzen and Thomas Graedel. Explains the greenhouse effect, long-term climate evolution and man-made effects.

The Rough Guide To Weather Robert Henson (Rough Guides, 2002)
Written by a scientist in the field, very clear and thorough. Includes overviews of climates around the world, the science behind the weather and hints on how to interpret weather forecasts and read the sky.

The Cloudspotter's Guide Gavin Pretor-Pinney (Sceptre, 2006)
Informative and enthusiastic, this is written by the founder of the Cloud Appreciation Society (www.cloudappreciationsociety.org). Few colour illustrations, however.

The liquid Earth

Mapping The Deep Robert Kunzig (Sort of Books, 2000)
In this compelling history of deep ocean exploration, Kunzig recounts the heroic endeavours of scientists past and present and describes the complex ecology of the deep ocean.

Disconnected Rivers: Linking Rivers To Landscapes Ellen Wohl (Yale University Press, 2004)
A survey of the environmental crisis gripping US rivers. Describes the impact of human activities on the US's rivers and related ecosystems, and reports on recent conservation efforts.

The icy Earth

The Ice Museum: In Search Of The Lost Land Of Thule Joanna Kavenna (Penguin, 2005)
In this blend of travelogue, reportage and literary essay, Kavenna recounts her search for the Atlantis of the Arctic. This is an endearing look at what the North Pole and its environs have meant to a succession of cultures.

Frozen Earth: The Once And Future Story Of Ice Ages Doug Macdougall (University of California Press, 1994)
The polar regions have given us many epics of exploration but few good science books. This book does much to close the gap. It charts the development of our understanding of ice ages, describing how they have shaped the landscape and influenced the course of human evolution.

The Earth and us

Guns, Germs And Steel: A Short History Of Everybody For The Last 13,000 Years Jared Diamond (Vintage, 1998)
In his Pulitzer-Prize-winning essay on human and environmental interaction Diamond argues convincingly that the reason Europeans have been dominant in world affairs for so long is the region's geography. Its suitability for farming led in turn to the domestication of animals, population growth, communication and innovation.

Gaia: A New Look At Life On Earth James Lovelock (Oxford University Press, 1979)
Lovelock's first book on Gaia, this has become a classic and is the ideal introduction to the theory. Also worth a look are *The Ages Of Gaia* (Norton, 1988), which charts the progress of Gaia over 4 billion years, and his latest, *The Revenge Of Gaia* (Allen Lane, 2006), which sounds a rallying call to lessen our impact on the planet before it is too late.

The End Of Nature Bill McKibben (Bloomsbury, 2003)
In this miserabilist bestseller, McKibben presents a frightening picture, backed up by the latest scientific evidence, of the destruction humans have wrought on the natural world. He has a deeper

point to make, however: as humans have become inescapable, we have lost something profoundly important – "nature" as a force independent and larger than ourselves.

The Unending Frontier: An Environmental History Of The Early Modern World John F. Richards (University of California Press, 2005)
A sweeping, 682-page history of human expansion and its effects on land and sea. Covering the period from 1500 to 1800, this book reminds us that we were damaging the world around us long before the invention of the combustion engine.

The Control Of Nature John McPhee (Farrar Straus and Giroux, 1989)
Three essays by the great geology writer on the human struggle against nature. Anything else he pens in book form or for *The New Yorker* is a must.

Climate change

Climate, History And The Modern World H.H. Lamb (2nd edn, Routledge, 1995)
The big picture look at pre-human-induced climate change. Full of good sense, this explains what we know about climate, how the past record of climate can be reconstructed, the causes of climatic variation, and its impact on human affairs now and in the historical and prehistoric past.

The Rough Guide To Climate Change Robert Henson (Rough Guides, 2006)
A clear, accessible guide to both the science of climate change and the political debate surrounding it. Including the latest research, it presents everything in an even-handed manner, giving readers a solid understanding of the issue.

A word on atlases

Wonderful things, computers, but a paper atlas of the world is one of the most satisfying objects you can own. I favour the ones with lots of maps, not page after page of graphics on coal production or disposable income per capita. While writing this book I had the *Dorling Kindersley Essential Atlas Of The World* on my desk and both the Concise and the Comprehensive versions of the *Times Atlas Of The World* close by.

Websites

One very well-known search engine which I often use has just produced 799 million references to the Earth. This is not a problem, because nobody would ever perform such a stupid web search unless they were out to make a point.

But beware. Any online search on topics such as fossils, geology and the like, or in areas such as cosmology and the age of the universe, will bring you straight to a raft of creationist websites, many of whose, er, creators have gone to some trouble to get a prime rank in web searches. Learn to shun them.

Many websites have been listed where appropriate throughout the book. A few of more general interest follow.

The solid Earth

US Geological Survey www.usgs.gov
Many thousands of pages of material on anything from volcanoes to coral reefs.

NASA www.nasa.gov
News and images from current missions, plus comprehensive information on the solar system and beyond. Check out visibleearth.nasa.gov for satellite images of the Earth.

European Space Agency www.esa.int
More spectacular satellite imagery of the Earth. The Reference section of the Space Science pages includes a wealth of information about everything from eclipses to detecting extrasolar planets.

Palaeos Timescale www.palaeos.com/timescale/default.htm
A detailed geological column, plus pages on each of the geological eras charting the history of life on Earth.

Paleomap Project www.scotese.com
A series of maps illustrating the movement of the Earth's plates over the last 1100 million years. Also includes maps showing the Earth's climate history.

The atmosphere and oceans

National Oceanic and Atmospheric Administration www.noaa.gov
A wealth of information on weather and the oceans. Includes the National Hurricane Center which gives detailed information about hurricane threats and the science behind them.

Natural Environment Research Council www.nerc.ac.uk
Includes links to its research centres. It is also worth downloading the leaflet *Climate Change: Scientific Certainties And Uncertainties*.

World Meteorological Organization www.wmo.ch
The UN arm responsible for promoting understanding of weather, climate and the oceans, to manage water resources and avoid the effects of natural disasters. Includes a wide range of material on their activities.

ESPERE (Environmental Science Published for Everybody Round the Earth) www.atmosphere.mpg.de/enid/English/
Click on "Climate Encyclopedia" on the left of the screen to learn about everything from the composition of the upper atmosphere to air pollution in cities and the El Niño effect.

Aviation Meteorology for Australia www.auf.asn.au/meteorology
An in-depth account of atmospheric structure and the physical laws which lie behind weather phenomena. Written for pilots, but of interest to the general reader, particularly those of a mathematical bent.

The icy Earth

British Antarctic Survey www.antarctica.ac.uk
Information about BAS's activities on Antarctica, including diaries from the research stations and vessels and a history of Britain's presence on the continent. Also includes a wealth of general information about Antarctica, its islands, weather and wildlife.

A Tour of the Cryosphere: Earth's Frozen Assets dsc.discovery.com/news/media/frozenearthvideo.html
This NASA video uses state-of-the-art animation and satellite imagery to take you on a tour of the icy Earth, highlighting the threats posed to it by climate change.

The Earth and Us

International Year of Planet Earth
www.yearofplanetearth.org
Information about the International Year of Planet Earth, scheduled for 2008. For short introductions to key sustainability issues, download "Groundwater – Reservoir For A Thirsty Planet?" from www.yearofplanetearth.org/downloads/groundwater.pdf and "Megacities – Our Global Urban Future" from www.yearofplanetearth.org/downloads/megacities.pdf

Museums

This book is virtually a catalogue of places to visit. But to gain some Earth wisdom, it might be a good start to take in an Earth-related museum. I live in the city that houses the Natural History Museum, but there are plenty of others, large and small, around the world. Here are some of the best:

UK

Natural History Museum London
www.nhm.ac.uk
Deservedly popular, the Natural History Museum hosts a wealth of permanent and temporary exhibits. Highlights include "The Power Within", where you can enter an earthquake simulator and watch terrifying footage of flowing ash and lava. Free entry.

Sedgwick Museum Cambridge
www.sedgwickmuseum.org
Smaller collections such as this are often at least as much fun as the big museums. As well as a good range of Earth science exhibits, the Sedgwick includes local displays on the geology of the surrounding fens.

US

American Museum of Natural History New York
www.amnh.org
The famous, huge fossil halls are informative as well as spectacular, while the recently opened Rose Center includes sections on the universe and on the solid Earth.

Field Museum Chicago
www.fieldmuseum.org
The core of the Field Museum's exhibits is its series of biological and anthropological collections, but it is also good on rocks and fossils, and its "Evolving Planet" exhibition is particularly good, charting 4 billion years of life on Earth.

Smithsonian National Museum of Natural History Washington DC
www.mnh.si.edu
Part of the Smithsonian Institution, this includes a broad selection of temporary and permanent exhibitions, with free admission.

Canada

Canadian Museum of Nature Ottawa
www.nature.ca
A good selection of exhibitions. The Talisman Energy Fossil Gallery explores the dramatic events that led to the extinction of the dinosaurs and the rise of mammals.

Australia

The Australian Museum Sydney
www.amonline.net.au
Includes "Biodiversity: Life Supports Life", an exhibit on unique Australian ecosystems, as well as the Chapman Collection, a beautifully displayed treasure trove of minerals.

New Zealand

Auckland Museum Auckland
www.aucklandmuseum.com
The Natural History section of the Auckland Museum explores New Zealand's unique evolutionary history. In the oceans section there is a striking life-size replica of the sea shore.

Photo credits

In text

15 NASA, Johnson Space Center; 21 NASA Jet Propulsion Laboratory; 25 NASA Jet Propulsion Laboratory; 32 ESA & NASA; 34 Bettmann/Corbis; 39 Richard Hamilton Smith/Corbis; 53 David J. Roddy/USGS; 63 Tony Waltham; 75 Tony Waltham; 86 Tony Waltham; 95 Tony Waltham; 114 (top) 18th Air Base Photo Laboratory, Hawaii/USGS; 114 (bottom) Kristi J. Black/Corbis; 115 USGS; 117 USGS; 130 Wolfgang Kaehler/Corbis; 132 NASA; 135 Robert Weight, Ecoscene/Corbis; 145 NOAA; 152 Jacques Descloitres, MODIS Rapid Response Team, NASA/GSFC; 181 Jacques Descloitres, MODIS Rapid Response Team, NASA/GSFC; 186 Adam Woolfitt/Corbis; 191 Macduff Everton/Corbis; 199 Nik Wheeler/Corbis; 208 Paul A. Souders/Corbis; 209 NOAA; 215 Roger Ressmeyer/Corbis; 221 Tony Waltham; 233 NASA/GSFC Scientific Visualization Studio and USGS; 243 Robert Simmon, NASA/GSFC/METI/ERSDAC/JAROS and ASTER Science Team; 253 Smiley N. Pool, Dallas Morning News/Corbis; 257 NASA Glenn Research Center

Colour section (each page clockwise from top left)

1 NASA Jet Propulsion Laboratory, Institute for Solar Physics at the Royal Swedish Academy of Sciences, Corbis; 2 Tony Waltham, Tony Waltham, Tracy Hopkins, Tony Waltham, Ron Watts/Corbis; 3 Michele Westmorland/Corbis, Sean White/Design Pics/Corbis, Frans Lanting/Corbis; 4 Tony Waltham, Tony Waltham, NASA, Tom Bean/Corbis; 5 Jim Sugar/Corbis, NOAA, Paul A. Souders/Corbis; 6 NASA/USGS EROS Data Center, Mark Karrass/Corbis; 7 Tony Waltham, NOAA, Tony Waltham; 8 James Leynse/Corbis, Keren Su/Corbis, NASA/GSFC/METI/ERSDAC/JAROS and ASTER Science Team

Index

A

B

C

D

E

F

G

H

I

N

O

P

Q

R

S